Understanding BIM

Understanding BIM presents the story of Building Information Modelling, an ever evolving and disruptive technology that has transformed the methodologies of the global construction industry. Written by the 2016 Prince Philip Gold Medal winner, Jonathan Ingram, it provides an in-depth understanding of BIM technologies, the business and organizational issues associated with its implementation, and the profound advantages its effective use can provide to a project team. Ingram, who pioneered the system heralding the BIM revolution, provides unrivalled access to case material and relevance to the current generation of BIM masters.

With hundreds of colour images and illustrations showing the breadth and power of BIM, the book covers:

- The history of BIM
- What BIM is in technical and practical terms
- How it changes the day to day working environment
- Why we need BIM and what problems it can solve
- Where BIM is headed, particularly with regards to AI, AR, VR and voice recognition
- International case studies from a range of disciplines including: architecture, construction management, and retail

Professionals and students in any field where the inter-disciplinary aspects of BIM are in operation will benefit from Ingram's insights. This book is an authoritative account of and reference on BIM for anyone wanting to understand its history, theory, application and potential future developments.

Known as the "Father of BIM", Jonathan Ingram designed and wrote the first Building Information Modelling systems. He has been honoured with the 2016 Prince Philip Gold Medal by the Royal Academy of Engineers for his "exceptional contribution to engineering" for "pioneering BIM". He taught the first courses in Object Modelling (BIM) at Harvard University, and was the winner of the 1990 British Computer Society Medal for Outstanding Innovation.

Understanding BIM

THE PAST, PRESENT AND FUTURE

Jonathan Ingram

Routledge
Taylor & Francis Group

LONDON AND NEW YORK

First published 2020
by Routledge
2 Park Square, Milton Park, Abingdon, Oxon OX14 4RN

and by Routledge
52 Vanderbilt Avenue, New York, NY 10017

Routledge is an imprint of the Taylor & Francis Group, an informa business

British Library Cataloguing-in-Publication Data
A catalogue record for this book is available from the British Library

Library of Congress Cataloging-in-Publication Data
A catalog record has been requested for this book

ISBN: 978-0-367-24413-2 (hbk)
ISBN: 978-0-367-24418-7 (pbk)
ISBN: 978-0-429-28230-0 (ebk)

Typeset in Calibri
by Jonathan Ingram

Publisher's Note
This book has been prepared from camera-ready copy provided by the author.

DEDICATION

To Harry, Alexander, and Maximilian, and, to Susana.

Editor's Note

I met Jonathan Ingram though the Autodesk Users Group International (AUGI) web-forum in 2015, while following a thread discussing the Origins of Revit. I was pleased to see that at least one other poster seemed to agree with me, that it was with SONATA, REfLEX and ProReflex, the last of which I had bought in 1997 from Parametric Technology Corporation (PTC) when Silicon Graphics Inc (SGI) released their O2 Irix 6.3 Workstation, the first BIM system available at a reasonable price, at least to sole practitioner Architects. It took a bit of effort to find him, having used a pseudonym for his post, but I wasn't surprised when he replied to mine and turned out to be the building software writer of our time.

I studied at Cambridge University School of Architecture (CU SoA) 1970-75 and knew many at the Centre for Land Use and Built Form Studies (CLUBFS), but it is was only much later through the internet that I discovered what they were all really doing in detail. Of course, we all bought their books and some of us attended the CU Engineering Department summer course on Computing, learning Fortran, preparing me for Telex machines, essential to UAE professional life until the early 1980s, when the fax machine was invented. I had also plotted the setting out of Bury Bus and Railway Interchange while working my 2 years pre-qualification for the RIBA with Essex Goodman and Suggitt, Architects in Manchester UK, using our Structural Engineer Bingham Blades' drafting system in 1977, prior to moving to the Midd e East. I can't imagine how much that table sized plotter must have cost at the time, but it must have been staggering by comparison with printing costs today.

CU SoA lectures and debate had convinced me that computers would eventually enable the whole building to be considered, including all its systems, perhaps even design and planning, where they had made a good start, not just lines and printing essential Drawings and Specifications required for daily professional practice. As anyone who has designed or built anything knows, it's the building cost that causes all the problems, often resulting from poor Bills of Quantity especially, so why anyone would have marketed a computer system that couldn't do it all is beyond me. I had considered the early systems and baulked at the price, so perhaps they were simply ahead of their time. A Whitechapel Computer Works MG-1 had a base price of around 10,000 pounds, half a Porsche 928S in 1984.

I have a chicken and egg theory about tools being accidental in their discovery, like using charcoal left over from cooking for cave painting, only to be developed once everyone has been convinced of their need, and hence the pencil thousands of years later. Does everyone remember adverts at the first computer shows c1980, suggesting that hundreds of thousands of dollars were required to make a computer drawing, compared with Michelangelo's masterpieces using a pencil? It was only much later that I discovered he didn't even have a pencil, but I expect that BIM only recently becoming widespread has nothing to do with its obvious benefits and desirability, but more to do with price, not the software but the computers on which to run it, and buildings if you don't. When I bought my first system, the software cost about the same, but the work station itself was much more expensive; over the last 10 years, all that has changed, but the cost and value of building continues to go up.

There is also the issue of knowing what you want before being able to design a suitable tool, often itself depending on knowing what's possible. I expect that the requirement for BIM was always there, but as the computers to process it were not, no-one knew they could have it, except of course, some of us lucky enough to have been at CU SoA when CAD was being invented.

It is an honour to be invited to edit Jonathan's book, having seen his name in documents enclosed with my PTC ProReflex CD so many years ago: better still to have been able to contribute to some ideas that will improve BIM for all of us, outlined in Chapter 17 concerning the use of Apps and Plug-ins, allowing standard office software to read from the BIM direct. Who would ever have guessed ?

Don't be in any doubt that without BIM, sole practitioner Architects are a thing of the past; everything will be designed and built by corporations, who can't do without it either. As for Clients, it goes without saying; BIM has it all.

Peter Dew MA(Cantab) RIBA

Table of Contents

Early Case Studies

Evolution of BIM

List of Figures

BRING YOUR BOOK TO LIFE
WITH AUGMENTED REALITY

GO TO YOUR APP STORE

DOWNLOAD THE APP **UnderstandingBIM**

RUN THE APP: POINT YOUR PHONE OR TABLET AT THE PAGES OF THE BOOK SHOWN IN THE INDEX, AND EXPERIENCE BIM IN GREATER DEPTH

Acknowledgements and Credits

I want to thank all those who supported me during the early years of developing SONATA and REfLEX and in particular my family and friends. A special thanks to Murray Pearson, whose practical input into the design, functionality and bug chasing helped make SONATA. My thanks also to Gerard Gartside, who played an important role in the development of REfLEX. My thanks to Bob Wakelam for his thoughts and memories of those times and to Nick Crane of BAM Construct for access to SONATA and his images and memories. Also, thanks to Pete Baxter for his support, suggestions and understanding over the years. I would also like to thank Peter Dew RIBA, Architect and regular ProReflex user until 2011, for his input. Also thanks to Rashmi Mistry for his efforts in sorting out the early versions of SONATA and to A.I. Grammarly for help with grammar and comma usage.

Thanks to George Stevenson for his help and images and to Ian Bevan, who was always a strong supporter of SONATA and REfLEX. My thanks also to the many other users who have been great supporters of SONATA and REfLEX, in particular Gerry Beers and Stuart Walker of Taylor Woodrow and also Mike Davis of Richard Rogers for his efforts before the main board of BAA. Also thanks to Chris Penn.

I would like to thank Jeff Findlay, Richard Vertigan, Giuliano Zampi and Martin Hall for the use of their images. Thanks also to Paul Wilkinson of pwcom.co.uk Ltd, to Stephen Davies, and to Maria Albanese of the Barr Smith Library at Adelaide University. Also, thanks to Charles Benson QC and to Thomas du Val de Beaulieu for legal and AI support respectively. My thanks to David Jellings for his in-depth information on BIM Objects and also to Shark Bay Productions and to Kim Neale for help with this book and its illustrations.

Others who have helped bring this project to fruition are Martyn Day of AEC Magazine and NXT BLD and, Ray Crotty of C3 Systems; my extended thanks to both of them for their support, help and encouragement in writing. Also thanks to Charles Eastman and Paul Teicholz for their review of an early copy of this manuscript.

Of particular note, I thank Mark Edwards (CEO of 345 Holdings) for his support in SONATA, REfLEX, in the production of this book and for the opportunity to immerse myself in the glorious new technologies, in particular, AI and AR, and to be able to explore Retail Information Modelling.

There are countless others who have help me along the way, please forgive me if I have not named you and my heartfelt thanks go to you as well.

The following companies and individuals have kindly provided drawings, illustrations and images for which I am grateful. If I have missed you and you belong on this list then thank you and I apologise. I recognise that some of these companies have either been acquired or are no longer in business but my gratitude extends to them.

CADAIM
Murray Pearson
Engineering Technologies
AutoDesk Inc.
Peddle Thorp Melbourne
Seifert Architects
Rosser and Russel
Mowlem (Carillion)
Phippen Randal and Parkes
Haden Young
Frank Shaw and Partners
Cadenza Designs ltd
Graphisoft
Taylor Woodrow
Kyle Stewart
Hickton Madeley
Alfred McAlpine
Sutherland Craig
Richard Rogers
BAM Construct
Norman Wright of SONATA Software Systems Inc
Unity Technologies
VergeVT
345 Holdings
PDA
PDP
Blee Ettwein Bridges Partnership
Leandra Boldrini and Pedro Oscar
Vectorworks

The Building Modelling Company
Seiferts
Sir Terrence Farrell
Mark Lester
Mark Edwards
Giuliano Zampi
Bob Wakelam
Scott Brownrigg (GMWP)
Julian Bricknell
Mike Letchford
Coles Australia
Massimo Guerini
Marks and Spencer
Mott MacDonald
Frank Shaw and Partners
SVM
BAM UK
Sensus Inc
Katerra Inc
Trimble
Unity
Architecture 2030
Boston Consulting Group
e_Arch
Victoria and Albert Museum
Richard Vertigen
Microsoft

Introduction

For most of my working life, I have saved things of interest into an old battered suitcase.

During the 1970s, my suitcase filled with 3D images, fonts, movies, and, in fact, anything graphical; there was no rhyme or reason, just items of interest.

During the 1980s, while writing SONATA, I continued throwing images, drawings, cuttings, documents and more movies into the suitcase. I found another case for the overflow.

While writing REfLEX in the 1990s, I threw even more of the same. By 1997 I was weary of writing systems and weary of throwing, so the suitcases were put away and forgotten.

Since then, I have watched, with interest, the up-take of building object modelling aka BIM. I had been somewhat surprised at the lack of clarity in the texts and by the lack of honesty about authorship and influence of these systems. In the end, the scale tipped, irritation overtook inertia, and I felt the time had come to retrieve the then 20kg of suitcases from storage and write this book. At the instigation of the Victoria and Albert Museum my battered suitcases found their RIBA Archive in South Kensington, London.

Over the years I have been often asked how I came to write such computer systems and how, decades on, I am still writing such systems.

As a young boy, I was raised in inland Australia, where by today's standards there was nothing to do; no TV and certainly no electronics. As a result I spent my time dismantling, designing and building, boats, planes, electric helicopters (powered from the mains), explosives, rockets, and war machines. My father had given me an account at the local hardware merchant. He stocked all the raw materials one could ever want; from sulphur and saltpetre to Uranium 237, from wood, nails, and glue, to electrical cable, motors, and magnets. I created many different projects and at one moment, I attempted to make nitroglycerin.

No one is more surprised than me, that I survived my childhood.

I went on to study Civil Engineering and, while studying, I was introduced to slide rules and computers. The former made me realize that there must be a better way and that the latter, was the better way (in spite of being fed with punch cards).

During this time, I designed a steel stadium, pipe networks, and other things that student engineers do. Everything was executed by hand using slide rule and pen and ink. When I graduated I worked as an engineer for a time, designing and setting out country roads. I used "chains" (a metal tape measure that needed to have temperature corrections applied), calculated by log tables and drawn by hand. Some of the resulting roads and small bridges still exist in remote parts of Victoria.

The pain of doing this by hand, believing there was a better way, led me to search for something. My first breakthroughs came when I worked as a scientist for the CSIRO, an Australian Government Research Organization. At the CSIRO, we had a very fancy range of computers, including the largest computer in the world at the time. These toys were an extension of my childhood, only more fun. I completed many different computer science projects: compilers, interpreters, databases, typesetting, and perhaps most importantly, a range of 3D systems. One of the more interesting projects was a hidden line movie of a street scene of buildings for planning approval for the Hobart courts. I still have the sprocketed 16mm film, possibly one of the first architectural movies generated for planning purposes. (To view this, download the UnderstandingBIM app from your app store). The tool set I developed was incorporated for the systems a few years later.

Some years later, I moved to England (via all the countries between) and there, knowing what I wanted to do, I found the perfect environment to hone my skills. I chose an architectural practice with heavy computer leanings, Gollins Melvin Ward Partnership. They were working with a 2½ D drafting system called RUCAPS (developed at Liverpool University) based on BDS concepts to which I added a separate stand-alone 3D modelling system.

The arrival of the first desktop workstations meant it was time to build what I had been formulating for some years; an accumulation of the systems from my days as a scientist and live architectural requirements. When GMW failed to see my vision I left, much to their annoyance. For my development I bought the first desktop machine available in the UK, a very expensive Whitechapel Computer Works MG-1.

Although I had a vision, it was only the beginning of several lonely years in my attic in Berkhamsted designing and building in what was to become the new CAD system (see Appendix 4 and 5). These are interesting in that they meet the criteria of a modern BIM system.

There were NO tools available to help; no libraries, no interfaces and almost no graphics capability. There was a Fortran compiler and a technique for setting pixels on or off on the monochrome screen. To do this, I had to set individual bits in memory to draw lines. I wanted the look of the new windows and pull-down menus, as recently developed by Rank Xerox. I had to write the overlaying windows system to manage the screen. Pull-down menus had to be done from first principles. The piece of code I was most proud of was a few lines of recursive code that managed the overlapping of all the different windows on the screen. It recursively divided each window so that only that part visible was drawn.

When the system was coming together, two years into the project, I was joined by Murray Pearson, an architectural draftsman. He was brilliant. We had only one machine between us. I would work on it from early till 6 pm, and he would do the night shift, often finishing at 3 am. He came up with lists of bugs and issues and I endeavoured to fix these in the day. Murray had considerable input into the final design of the system. Rashmi Mistry extended the team to three late in 1987. He helped hugely in sorting out the tangles.

The result of all this was SONATA.

I invested in a colour screen and went to the USA and showed it at a CAD conference (well actually a hotel room near the conference). The idea of having automated coordination of plan, elevation and 3D and the complex parametrics sparked interest in various companies. Autodesk expressed an interest and said they wanted it, and in October 1987 we went to Sausalito with high hopes and little cash. There we demonstrated the capabilities to senior members of Autodesk and started talking seriously. Halfway through the meeting the phone calls started. The bad news was that it was on Black Monday 1987 and we went home empty-handed. I do remember an employee's car in the Autodesk car park with license plate "AutoBad".

Despite my reservations, I was persuaded to sell SONATA to GMWC. It went on to be sold to the Canadian company Alias. Alias overstretched themselves and abandoned a number of projects including SONATA. There were distributors in 25 countries at one time.

Perhaps the first real user of SONATA in real projects was Mark Edwards, an architect and engineer. He built a team to fix the many bugs I managed to include. He is still putting teams together to fix my bugs.

In 1992 I decided to have another go. Technology had progressed leading to improvements in the concept, in particular the compilation of objects and merging with the system at runtime. This meant vast improvements in execution speed. Many graphical libraries were available simplifying the coding. I was in contact with some of the people who had worked on SONATA; Gerard Gartside joined me, and together, we built REfLEX Systems. Initially we worked in my house in Berkhamsted, but when we had 12 employees, my very patient ex-wife strongly encouraged us to move into an industrial unit up the road. Around that time I was appointed Professor of Engineering and Construction Management at the University of Reading, but was unable to take up the appointment because of the subsequent sale of REfLEX.

In 1996 I negotiated and sold REfLEX to Parametric Technology Corporation (PTC). Almost the whole company moved to the USA where we started work on ProReflex. There I was appointed Chief Technology Officer of PTC.

The founders of Revit, employees of PTC at this time, had been working on the mechanical CAD system Pro/ENGINEER. When they left PTC, they acquired a non-exclusive development rights ProReflex (REfLEX) including all code, documentation and access to the REfLEX developers. They went on to develop Revit; the overlap in functionality between Revit and REfLEX/ProReflex is huge. I still have Revit Version 1 from 2000 running on my desktop. Some of the errors between the two systems are similar and possibly identical (see the Understanding BIM app for developments). One of the claims on the Revit 1.0 box is "The first parametric building modeler"; there are at least two systems before Revit. The amount of accrued development over the years from software and usage must be in the thousands of man years if you go back to BDS, RUCAPS, SONATA and others. I find it hard to believe that the Revit developers arrived at this point without in-depth reference to ProReflex/REfLEX/SONATA, given the similarity in most of the functionality (except families).

I left PTC after several years to pursue my own interests. Over the following years, I founded startups, innovated, and patented in various fields, including sensors, electronics, renewable wave energy, pharmaceuticals, and tyre pressure measurement. I also taught postgraduate Architects at Harvard on Object Models in Architecture, I had a research project in AI with Prof Jean Jacques Slotine at MIT, I wrote a light opera (lyrics, libretto and music)(watch the billboards!), and several other projects. Somewhere in there, there was a book called "The Miracle Molecule". The molecule was miraculous as a cancer cure, but failed phase 3 trials with FDA due to a change in protocol.

It has been fascinating to watch the development of technology. When I started technology was a room full of punch card driven computers with the processing power found in a contemporary musical birthday card. Memory, speed and storage was measured in thousands and now everything is measured in billions.

More recently, I have returned to Information Modelling, applying the principles of BIM to Retail. This has been spiced up with these technologies of Augmented Reality, Artificial Intelligence, Virtual Reality, voice recognition, and language understanding, and, best of all, really fast machines and mobile computing.

The new app that comes with this book is an experiment in just this (see app store for UnderstandingBIM). The idea is to bring the book to life, with animations, models, videos, additional images, interviews, other media, and as yet untested, a direct social media link. Material can be added to the app even after this book is published.

In terms of man-machine interfaces and capabilities, Artificial Intelligence and Enhanced Reality take us to new realms. My belief is that one day soon, we will have systems that are cognisant, capable and coherent, aiding and assisting many aspects of human endeavour. I sincerely hope that my own efforts will inspire designers of the future.

Jonathan Ingram

London
2020

Addendum by Mark Edwards (CEO of 345 Holdings)

"My favourite story on that was being a naive architect just out of university I started using the first cut of SONATA on a live project (as you would) and I was working through the night on a deadline when I tried a Boolean operations on a curved wall and the whole project disappeared at 3am, I spent the next 5 hours panicking trying to recover as I had no idea what was happening.

When my boss came in he saw I looked exhausted and said ok, let's call Jonathan.

You came in and asked what I'd done, to which I replied 'what I've done, it's this bloody software '.

You said can you tell me exactly what you did in which order, which I did and you said aha that's a 'bug'...! What's a 'bug' I replied and you explained to which I innocently asked why you'd put that in there then.

You then asked what else I'd like in the system and I rattled off 5 ideas and you said 'I can have 4 of those by next week but the 5th will take longer' and that's exactly what started my career because when you delivered the new functionality I realised what was possible with the right minds and being able to talk each other's language and I've been doing it ever since...!"

Why Do We Need BIM? 1

An intelligent, computer-based model of a structure must help in its conception, design, construction and management. This process, now known as Building Information Modelling or BIM, has become the mainstay of modern building and infrastructure design, construction, and management. It forms the basis of a complete new digital approach across the Construction Industry, from town planning through structural analysis, to highway and tunnel design through ships. This book takes us through the journey of this revolution, how it was achieved, how it is used and where it is going.

This first chapter details the traditional drawing-based approach to building design, how this tends towards fragmenting the design process. It shows how BIM is necessary to solve this and other problems.

DRAWING BASED CONSTRUCTION

Building construction is one of the most information intensive industries. Buildings (and infrastructure) are complex and the projects behind their construction are equally complicated and ever-changing, invariably generating huge volumes of information which must be accessible to large numbers of diverse companies and individuals. Many of these are united for the first time by the project itself. Traditionally drawings have been used to represent the basis of the building. It is difficult to imagine some of the buildings of antiquity being built without some form of drawings.

The earliest surviving "construction drawing", a stone wall etched with the profiles of columns and mouldings, was found in the Temple of Apollo at Didyma constructed just before the time of Christ. Since then and probably before then, buildings, roads, harbours and mausoleums have been built using designs set out on paper, papyrus or in the case of Didyma, stone.

The very human skill of drawing has a history dating back tens of thousands of years, certainly before writing and possibly before language. It is one of the more important intellectual skills that our species has developed. The ability to capture an abstract representation of the physical world in a form that enables it to be communicated with others is a remarkable gift.

Drawing is a primary form of communication, where artists convey views of the world they see, but the meaning and intent of many drawings can be diverse and potentially unclear. In some cases, the viewer must spend time learning how artists approach their subjects, the techniques and conventions used to represent them. The viewer must make an effort to learn the explicit language of the artist.

These and many other issues around meaning or communications, come to bear, in one way or another, at the more mundane level of technical drawing. For us, a drawing is a two dimensional (2D) depiction of a physical object, typically created using lines or shading to represent the visible edges of the object. It is usually seen from a given view point, perhaps shaded or hatched to illustrate materials and 3D form.

Lines may form simple shapes such as polygons, ellipses and other conic sections. Such lines may readily be traced out using tools like templates. Alternatively, they may take the form of more complex shapes that can only be drawn freehand, or using a device like a spline tool, or by plotting, point-by-point, the path of a mathematical function. Such lines can be very difficult to generate precisely in the first place and difficult, if not impossible, to subsequently reproduce accurately.

To fully represent a simple three-dimensional object, it is usually necessary to develop separate views of the object in question. A single 3D drawing will certainly show the essence but to accurately show the complete object different views of the object must be formed. Each view in turn may contain a symbolic representation of the actual geometry. It might also contain other detail in order to show all facets of a complex object. Each of these views comprises its own array of lines of different types, including parts of lines where one edge of the object passes from one view to another. In many circumstances it is usual to apply additional drawing conventions such as colours, line-styles, hatching and shading to add realism or technical detail to drawings.

This is the sort of process that is involved in drawing a single, simple object. However, buildings usually involve a large number of highly complex objects connected in different ways. To show the whole structure uniquely requires many drawings, and even then aspects of the design may be hidden. This is greatly complicated by the need to accommodate different views for each of the many disciplines involved. Ensuring all of these drawings are correct within themselves and consistent with each other as the work evolves through various designs and development processes and iterations is a major task.

A drawing is a store of information, where the data are individual lines and text, each of whose attributes include its point of origin, its path and length. There is only so much information that can be gleaned from lines on a 2D page. Somewhat stating the obvious, our world is very much 3D but also varies in time.

Construction drawings, no matter how detailed you get, you can only say approximately what you mean or intend. We depend on the person at the other end of the conversation to complete the picture and read the drawing correctly, to understand it fully and use it intelligently in his or her particular situation. And no matter how much effort you put into it, your drawing will never be unambiguously clear, complete, correct, internally consistent and coordinated with other people's related documentation. With drawings, certainly large numbers of drawings, this is simply not possible.

As a result, huge numbers of people across the construction industry have spent inordinate amounts of time checking information, guessing the true intent, sometimes getting it wrong, correcting things, making mistakes, cutting stuff out... all because the only way we could design buildings was with drawings; drawings that are potentially untrustworthy, at odds with other drawings, unintelligent, in-computable information, at least some of the time.

When you have done this every day of your working life, or worked with other people who do so, it can be difficult to stand back and appreciate just how complex and inherently problematic the process is in real terms. Transferring the process

to a computer, using a computer-aided drafting (CAD) system, might appear to simplify things, and it does to the extent that the machine can be programmed to store all of the necessary information about each of the individual lines. The computer can ensure that lines that are supposed to intersect at a particular point do so, that lines that are supposed to be parallel remain parallel, and so on. Producing drawings on a 2D CAD system or even derived from a 3D CAD does not, however, guarantee that the drawings will be consistent.

A set of drawings produced on any CAD system looks convincing. They are drawn perfectly (compared to manually produced drawings), parts are replicated precisely and the patterns and text look printed. However, the real issues are hidden. If the drawings are produced in a conventional CAD system as separate plans and elevations, then the chances of them being perfectly coordinated, matching each other exactly when detailing the same items in the building, are minimal; the bigger and more complex the building, the greater the risk of error. The larger the number of documents produced, the greater the likelihood of error, and, the lack of coordination between the drawings and the ambiguities.

Related problems derive from the fact that CAD operators produce large numbers of drawings – just because they can. Designers, swamped with stacks of convincing looking drawings, become over-confident, often neglecting to check details, particularly at interfaces between different elements and disciplines. When teams come under schedule or production pressure and have to recruit additional staff, the newcomers usually have to get up to speed very rapidly, potentially compromising existing quality and technical standards – all without seeming to do so.

Compounding the possibility for error, information brought together to complete the design of a building usually comes from diverse sources. Professionals from different disciplines each contribute in varying ways to aspects of the design at different times. The architect provides the overall concept in terms of base drawings, possibly a 3D sketch; various engineers contribute their specialist knowledge; construction managers, town planners, lighting specialists and the client or end-user all contribute and comment. Coordination of all this input is problematic. There are bound to be conflicts between the various types of information, even about what is intended, because there is no unifying single reference point. Even within single organisations, information is often fragmented.

Finding a common base where this diverse information can be stored and compared is also problematic. The diverse professionals tend to store data in formats specific to them, making comparison difficult if not impossible.

Traditional CAD programs do not assist in this coordination either. The needs of the different disciplines are diverse, requiring information in different formats and outputting very specific data, usually in highly specified formats. For example, designing steelwork or laying out building services are specialist applications, almost always with their own databases and at best generating sets of 2D drawings that represent something in 3D. Visualizing and coordinating this information is again fraught with pitfalls; the framework is often fluid, making true comparisons impossible.

Buildability of the structure is an additional factor that is not addressed with these types of representations. Can the steelwork be assembled; can the lift be brought on site and put in place? Buildability must be addressed at an early stage, otherwise different contributions from different disciplines may result in an assembly process that does not work in practice.

The structural engineer changing beam size, services engineers changing ductwork size and position, perhaps even a simple change of room layout can all lead to potential clashes of two or more objects trying to occupy the same space at the same time. Discovering and resolving these problems on site is an expensive and unsatisfactory solution.

These various issues of coordination, design, buildability and sharing information across disciplines were not easily solved by conventional drawings, nor by traditional CAD systems.

The information needed by the different disciplines is directly related to what they need to do. Much of the information is not graphical, but rather lists of items, costs, delivery schedules, stresses and strains, amongst other things.

This non-graphical data was usually input in a form suitable for the particular systems being used to specify that aspect of the building. Some of this information is required by several

Figure 1. *The problems of building a swing*

disciplines. Sharing the information had to be done explicitly, outside the drawings. This was, of course, problematic and led to further lack of coordination and consistency.

Some stakeholders in a project require a 3D view of the structure. For instance the town planner or Joe Citizen needs this type of view. Coordinating the 3D model with the different 2D views of the structure has always been problematic. Some early CAD systems have allowed the projection of 3D onto a 2D drawing and then allow annotation of that drawing model. This solves the coordination problem temporarily but as soon as there are changes, and there will be, the 2D drawings are out of date. This was later referred to as 3D+2D as 3D projections cannot be used direct as 2D drawings because many of the symbols used are symbolic, needing particular representation.

Another issue comes from the level of detail required at the different stages of design. The town planner does not want to see reinforcement detail and a massing diagram is insufficient for the facilities manager. Similarly as the design develops the level of detail and its accuracy increase.

Construction managers also need to see the structure in a temporal sense; how the building will look at different stages of construction. Overall planning, coordination and control of a building project from start to handover requires time-based information. The problem of what and when items need to be ordered, delivered, constructed and maintained is very different from the straight analysis.

This information needs to be related to each of the sets of drawings and models coming from the development team, where integrating the construction sequence into the design brings significant benefits. The allocation of jobs and the sequence in which particular construction tasks are to be initiated becomes of paramount importance when building any structure. 3D drawings do not really tackle this 'temporal' or programming problem, traditionally tackled by a range of project planning software systems, which needs to be integrated into the overall design of the building.

Another issue faced by the different stakeholders contributing to the design of buildings was that they were often dispersed throughout different offices and locations, possibly around the world. Being able to interact with other professionals while designing (and constructing) a building is important to achieve the most effective design, but may not always be easily managed. Simply sending a set of drawings does not quite work, and is anything but conducive to collaboration.

The Construction Industry has struggled with these problems in building since the splitting of the different design aspects of a building. Formerly solved by the Master Builder, the on-site expert of ancient times, these issues have become more and more predominant with the increasing complexity of buildings, specialist technologies and complexities of the sub disciplines.

Figure 2. *Drawing Coordination was problematic*

VIRTUAL BUILDING MODEL

Building Information Modelling (BIM) proposes to solve all of these problems. It does so by building a virtual or computer based prototype of the structure using a technology that enforces coordination, encourages sharing and allows and in some many ways enforces design consistency.

Such a system needed to be an object based information model, representing the building in all its particularities. All information had to come from that single model. The different disciplines involved needed to be able to work concurrently to build their parts of that model across all 2D, 3D and other views. An internal understanding of connectivity and to some extent, function must be inherent, together with automatic detailing and design of the different parts.

Consistent databases cannot have duplicate information representing the same thing, if one is changed that consistency will be lost. All information must be defined once, and referred to at other points where that is information is needed. Whether inside or outside the model, all information should be defined once. When information is changed then all downstream related information must be updated.

BIM – CONSTRUCTION WITH (EFFECTIVELY) PERFECT INFORMATION

In general, design information generated using an object or information model-based process is inherently superior to comparable information generated using drawing processes[1]. In BIM, this is particularly true. Moving from CAD to BIM is not simply a matter of upgrading from one CAD system to a better one. In moving from CAD to BIM, we are introducing a truly fundamental change, a complete change in the nature and quality of information we work with and radically transforming how the entire construction industry works. It also introduces, necessarily, a change in work practices.

Put simply, but certainly not exhaustively, BIM involves the creation of a virtual 3D building model using not lines on sheets of paper or computer screens, but complex, fully specified, intelligent parametric objects, in a digital, multi-dimensional environment. These objects or components can be as detailed and as accurate as needed. They appear to behave just like building components in the real world. Doors, windows, walls, beams, pumps, luminaires, etc., all display properties in the modelling

system which correspond exactly with those of their counterparts in the real world. Individual models, or parts of models created in this way, are inherently accurate and complete. They can contain extremely high quality information, rich with embodied intelligence and other forms of programmed behaviour.

What would happen if, instead of drawings, the information used in construction were fully trustworthy – needing no checking – and was readily computable, as the BIM vision promises? What would happen if the operation of the construction industry was based on the use of effectively perfect information? A few things come to mind.

One significant problem with our dependence on drawings, is their use as the basis for the procurement of traditional construction contracts. A minority of clients already procure in more progressive ways, such as framework agreements or alliances, with bidders evaluated on, say, 'best value' rather than lowest price, but most do not.

Effective competition in any market requires that the customer can specify their requirements accurately and in such a manner that competing suppliers' proposals can be compared and evaluated transparently, on a true, accurate, like-for-like basis. This is almost impossible to achieve using drawing-based documentation, where the scope of work can often be interpreted to mean almost anything a bidder can plausibly claim it means – which means that there is no definitive scope of work under the contract, certainly not one that can be challenged.

This in turn means that it is impossible to eliminate predatory bidders from the contracting process, as all bidders for a given contract know that one of their competitors may adopt a "bid low / claim high" strategy. They must all therefore bid as low as they dare, and hope to make their profit on re-interpreted work, claims and other extras. This behaviour eliminates the possibility of real competition for the operational components of traditionally procured construction contracts. Competition among contractors today is mainly about the marketing and estimating skills, the commercial nerve required to win work, and the claims management skills required to make money from projects won at cost or less. Skills in construction operations may give project teams a sense of pride and achievement, but are largely irrelevant to the survival of the firms they work for.

So contractors have no imperative to innovate, and avoid innovation risk and investing in improved production methods. Instead they sub-contract, and sub-sub-contract, right down the supply chain to the point where subsistence level, labour-only subcontractors, working in gang size firms, perform the bulk of the

---

1   "The Impact of Building Information Modelling: Transforming Construction", Ray Crotty, Routledge, 2011

industry's work. These organisations have neither the vision nor the wherewithal to invest in improving project delivery processes.

As a result, in the pre-BIM world, effective competition can be said to exist mainly at the top and bottom ends of the construction industry: competition of ideas among designers and product competition among manufacturers. Nearly everyone in between competes to win projects – not to deliver them: a crucial, crippling distinction. Alternatively, designers are tempted to insist on particular products being used rather than draw and specify them, where the allegation of corruption isn't easy to avoid.

Trustworthy, computable tender documents would help transform this situation. With near perfect, complete scope definition, bidders are compelled to compete on the basis of their ability to perform the construction work efficiently. Every line item can be linked directly with a component in the model and must therefore be priced explicitly. Every price can be compared automatically and challenged as appropriate. There would be no claims opportunities except for changes post contract, so bidders must get it right going in.

As in other efficient markets, competition would force contractors to improve these techniques continuously. Efficient firms will profit greatly – they will no longer be undercut by claims-hunters. Profit will no longer be squeezed out of the industry. Construction as a whole will become wealthier, able at last to invest seriously in people, methods and physical capital; labour productivity will soar.

MANUFACTURED BUILDINGS

The precision and computability of model based designs enable physical components of buildings to be machine-made directly, using the data contained in the modelling systems (as happens today with Frank Gehry's buildings[2]). The idea of 'tolerance' will disappear; individual objects will be manufactured with near perfect precision (where this is appropriate) and will be pre-assembled, equally precisely, in the factory, before being shipped to site. Individual components and sub-assemblies will be dropped, clipped, or slotted into place, using the sort of 'click-lock' coupling techniques found in electronics and other areas of manufacturing. The key point is that no manufacturing from raw materials and no shaping operations – no pouring, cutting, routing, drilling, bending, folding of raw components – will take place on site. The site becomes an assembly plant – comparable to a car assembly plant. This will be a craft-free industry, not through

lack of craftsmen as is often the complaint regarding quality today, but through lack of need of them.

It will also be a super-fast-track industry. Knowing that the other elements of the building have been (can only be) assembled exactly as designed, means that, instead of having to wait to check whether earlier elements have been built correctly, the manufacture of all components could, if required, commence simultaneously and proceed in parallel, as soon as the model has been completed. This has many advantages, including competitive sourcing from whichever part of the world is most appropriate, and hugely reducing the time required for site assembly, or construction.

GUARANTEED BUILDINGS

Just as long guarantees are an important attraction to buyers of motor cars and other complex products, so long guarantees are likely to help drive the market for buildings of the future. Suppliers will emerge who will offer, say, 20-year guarantees covering all of the performance characteristics of a building that can be simulated and tested in a BIM model. This will include the maintenance performance of the fabric of the building and of the equipment within it, the building's energy performance and even the ease with which it can be re-configureured for new uses during its lifetime. Suppliers of these buildings, for example, Rolls Royce with its aero-engines, or closer to home, Otis Elevator Company, will aim to derive as much of their revenues from servicing the product in its life in use, as from the initial sale. This mode of operation will ensure that buildings in the future are designed and built to optimise their whole life costs. It will also require that performance feedback loops become an integral part of their operation and maintenance, ensuring that their suppliers become real learning organisations, with an inescapable commitment to the on-going service of their products.

Links between main contractors, or assemblers and component manufacturers, shall be dramatically shortened. Specialist subcontractors may continue, but only as part of larger contracting organisations, or as licensed and thus guaranteed installers of manufactured equipment or systems.

Major equipment and building systems manufacturers (GE, United Technologies, Siemens, Permasteelisa...) could become the main contractors. They already know how product design, component sourcing and assembly processes work, as this is how they currently deal with their mainstream products: lifts, cladding systems, switchgear, mechanical equipment etc. Some of these firms will surely extend the range of their activities, perhaps

2   "The Impact of Building Information Modelling: Transforming Construction", Ray Crotty, Routledge, 2011

by acquiring construction firms, to include the production and servicing of complete, guaranteed buildings.[3]

MANAGED BUILDINGS

Having built a comprehensive model of the building, with all its complexities, we are well placed to use the model to manage that building into the future. Maintenance, performance, access, and complete drawing sets can be all included in the model.[4] Asset or facility managers in particular would benefit from using such a model owing to the following:

- The asset model is fully populated with references;
- Assets within the model are capable of, and should be, self-documenting, or at least direct references to external files;
- At handover, all documents can be referenced from a single platform;
- Operations can be streamlined beyond the initial project phase;
- Issues with particular assets can be determined quickly with access, other asset proximity and so on being obvious outcomes;
- It is probable that assets may have the ability to be monitored from within the model;
- Life-cycle costs can be determined from within the total environment;
- Assets can be pro-active in terms of announcing they need service or monitoring;
- Collaborative working is possible through the shared model and hence the shared asset model;
- Commissioning, training and handover can be achieved more quickly with fixed, guaranteed correct, data optimizing operational performance.

---

3 "Construction 2025", HM Government, 2 July 2013
4 "Government Soft Landings", BIM Task Group HM Government

TRANSFORMED CONSTRUCTION – CHANGING ROLES

In the scenario envisaged here, buildings become products; huge, complex products admittedly, but products nonetheless. The market for buildings will be characterised by powerful, effective competitive forces that will compel their suppliers to innovate continuously. The future we face is one of effectively perfect information, together with almost unlimited computing power. As a result, it is likely that anything – any and all human activity – that can be programmed, will be programmed – embedded in the hardware or software of computerised systems. Most other engineering-based, rules-based, industries have already been changed out of all recognition by this process of digitisation. Construction is now undergoing a similar transformation.

BIM can address all of these problems by ensuring a single coordinated intelligent model, enforcing sharing of the model across the different disciplines and users, and providing a temporal and live model representing the state of construction.

Figure 3. *SONATA Advertisement © Sonata Software Systems 1988*

What Is BIM?

2

BIM is seen as one of the greatest technological innovations in the construction industry.[1] Most larger buildings around the world are designed and built using BIM with over $1 trillion expected in annual savings world-wide by 2025.[2] Information Modelling can be found through the design and construction of buildings and has spread into other areas including city and infrastructure design and management. It has brought change across all disciplines associated with the construction industry; governments around the world have legislated for its use.

What is it that drives this uptake? It solves problems and saves time and money, it brings control and certainty, and it ensures coordination of design.

BIM centers around constructing an intelligent model shared across team members. This model acts as a collaborative hub throughout the entire life-cycle of an asset. The model usually starts at asset conception and gains intelligence and content as it gets passed between the different personnel involved in the design process. They add their expertise, document conditions, or add operations and maintenance actions.

This model allows design and other information to be extracted for all possible needs throughout the whole life cycle. It is made up of intelligent coordinated parametric objects that can be changed and manipulated yet are guaranteed to remain consistent and correct, regardless of how many times the design changes. The data remains consistent, coordinated and accurate.

BIM is not just about the technology though, it is concerned with work practices and in fact, engenders new practices. These work

practice changes are effectively "enforced" by the BIM technology. It has been stated that BIM is the process and not the technology, but one without the other is not effective.

In this chapter we introduce the underlying technology in simple terms, and why each part is necessary to achieve BIM's broad benefit.

SIMPLE OUTLINE OF TECHNICAL BIM

As stated, BIM involves the assembly of a computer-based model of the building from parametric component parts. This model is used to generate all construction and design information.

There is only one BIM model for each structure being designed with the different disciplines sharing that model. This is done to avoid duplicating information, to ensure consistency and coordination thereby avoiding the possibility of different 'conflicting' information. The BIM model is assembled from components that represent parts of the building; for example, walls, floor slabs or windows. Information about the objects making up the model may be referenced by the objects to files across the Internet. These are known as BIM objects and are readily available across the Internet.

Construction information is extracted from the model in terms of drawings or other information. Drawings are not saved as such, but are extracted views of the model and are regenerated each time the drawing is required. This is also true of all interactive work, all screen views are updated every time there is a change. One of the requirements of BIM is that a change anywhere in the model or related data updates everything correctly across the whole model.

Another important aspect of BIM technology is that the components making up the model can draw themselves differently depending on how they are being viewed. For instance, a light switch is drawn symbolically in plan, a different symbol in elevation

---

1 "The Driving Force of Government in Promoting BIM Implementation", UK Government, 2015

2 "The Transformative Power of Building Information Modeling", Boston Consulting Group, 2016

and an actual drawing of the light switch in 3D. It might also be drawn as a carefully crafted, visually accurate model for the computer generated image (CGI). It may have a functional switch that actually operates a light or lock. In SONATA in the late 1980s, the Author operated a doorbell and opened the door using switches built into the different view of the door.

Traditionally a door also appears differently in the different views (Figure 4). For each item the appropriate view appears in the current drawing or screen. Obviously each view needs to be defined for each component before this can happen. Such definitions will include sectional information and usually other information such as delivery, construction date and so on.

Object definitions may be defined as fixed graphic definitions or, as variable parametric definitions. As an extension of the door definition, it may draw itself in different configurations; as a single door or a double door, and possibly a glass door with different ironmongery and so on. Different views of windows and doors inserted into walls are shown in Figures 7 and 9. The default choice of detail about the door is usually made when the door is assembled into the model, but it might be edited to include changes required at a later date. Other information such as supplier, delivery dates, thermal properties and so on will be associated with that door. The certainty of the information associated with a component is one of the issues in developing a model.

The views within a single component must be coordinated between themselves. Revit's and ArchiCad's graphical parametric generator does this well and avoids a lot of the issues associated with manual techniques. Manufacturers and companies have generated many thousands of such BIM Objects with comprehensive detail. They do this to make it easy to specify their materials and goods. These can be found on websites such as BIM Object[3] and BIM Components[4] and so on. With large libraries available it is less likely that one will be defining components.

BIM technology allows Information in the model to be passed between different items that are adjacent or linked. For instance, a wall with a door "embedded in it" will "ask" the door for dimensions so it can adjust the cavity detail, adjust the brick count for the quantity surveyor, adjust the graphics in 3D so the gap can be seen through if the door is open and so on. "Ask" means ask invisibly within the technology. A link occurs between the door and wall, and information is passed automatically via that link. Door frames, the amount of plaster and paint, etc., all must

reflect the size of the door. In terms of windows and doors, the size of the hole will depend upon the parameters of the frame. If it's a double door of a given size, the wall must be able to extract that information and draw itself accordingly. To avoid duplicating information, data is passed between the door and the wall automatically, hidden from the architect's view, and is part of the software. If you allow the wall thickness or door size to be stored in different places and one value is changed, the model is no longer consistent.

The "ask" is not limited to doors, windows and walls, but is found throughout the design. Similarly walls might "talk" to walls to show the correct connection types, i.e. cavity termination and so on. Without a change in the system this will not propagate through leaving out of date information. At a more advanced level, further fluid rates, bending moments, loads, sizing, heat flow, electricity flow and other information can be passed around networks of objects and solving issues within the model itself. If you look closely at all the drawings in this book, you will see all the cavities, joints, and closures are completed.

The cavity termination and wall joints was a difficult problem to solve. Walls of varying thickness and cavity detail must match graphically in the symbolic views as well as the 3D.

Each time a drawing is generated or a view is changed on the screen the view is regenerated, reflecting the latest data. If there has been a change in the data all views will change appropriately. In a true BIM system it is impossible to generate data that is not coordinated and up to date.

Although not perhaps totally obvious yet, many of the problems from Chapter 1 have already been resolved through using a single parametric model, where information is passed between component objects.

FORMAL BIM DEFINITION

The National BIM Standard of the USA (NBIMS) definition of BIM is *"a computable representation of all physical and functional characteristics of a facility."*[5] Although this ignores some of the wider implications of BIM, as a technical definition, it is a precise description of BIM. In more formal terms:

Building Information Modelling is a digital representation of physical and functional characteristics of a facility. *"A building information model is a shared knowledge resource for information about a facility. It forms a reliable basis for decisions during the*

3   https://www.bimobject.com/

4   https://bimcomponents.com/

5   https://www.nationalbimstandard.org/

buildings life-cycle; defined as existing from earliest conception to demolition", as defined by the US NBIMS and in the UK by the National Building Information Model Standard Project Committee (UK). It involves the creation and use of a comprehensive building model to resolve, inform and communicate project decisions. Design, visualization, simulation and documentation are achieved within the model. Fully coordinated information and documentation are guaranteed.

In the "BIM Handbook"[6], BIM is defined as:

"Building Information Models are characterized by:

- *Building components that are represented with intelligent digital representations (objects) that 'know' what they are, and can be associated with computable graphic and data attributes and parametric rules.*
- *Components that include data that describe how they behave, as needed for analyses and work processes, e.g., takeoff, specification, and energy analysis.*
- *Consistent and non-redundant data such that changes to component data are represented in all views of the component. Coordinated data such that all views of a model are represented in a coordinated way. "*

EVOLUTION OF BIM NAME

The term Building Information Model, was first used in a 1992 paper by G.A. van Nederveen and F.P. Tolman[7], and has become, through mutual agreement of the main vendors and Jerry Laiserin's famous 2003 BIM debate[8], the standard way of referring to this concept.

Building Modelling, as distinct from Building Information Modelling, has been in use since the 1960s and was applied to early systems such as BDS, ARK-2 and CEDAR. (see Chapter 4). These systems were not BIM as we know it today, but modelled buildings with the intent of extracting certain information from the model.

From 1985 until 2003, the Author used the term Building Object Models referring to the same single model. This was chosen because of the similarity between BIM objects and C++ Objects in Object Orientated Programming, complete self contained

containers of different functions and their containing complete data. Various publications refer to the systems as such and the Author delivered lectures to Harvard Post Graduate Architectural students on this subject. The Author was invited and spoke at the South East Asia Regional Computer Confederation (SEARCC) in Sri Lanka. He presented a paper on *"Object Models in Engineering"* discussing the use of *"a new class of objects that have been developed to allow designers to prototype large engineering structures."*[9]

The term "Building Information Warehouse," coined in 1994 by George Stephenson of Engineering Technologies was used to refer to groups of BIM objects making up libraries.

CREATE DESIGN INFORMATION & RESOLVE DESIGN ISSUES

Stated differently, BIM systems allow the creation of component objects and enable their assembly into a building model. Once assembled, they can be seen in relation to each other and they can interact with each other and can be manipulated and altered individually or in groups.

The design assembly processes take place in different projections on the screen, though usually the plan predominates. It is easier to place things where you want them on the horizontal plane, with height being given by the datum of the layer or by an offset from that datum.

In an image from SONATA in 1986, Figure 36 shows multiple windows/projections that look into a building model. If you look carefully you can see symbolic and 3D views of the same components in the different views. The sectioned objects are automatically hatched appropriately. The objects only occur once in the model but represent themselves differently depending on the view (and the scale). Moving or changing a component in any window, causes it to move or change in the model. Hence all windows, information and drawings are updated automatically. This is automatic and needs no intervention. The information on the screen is identical to the information plotted in drawings. This is again a technical characteristic of BIM systems.

Every component has at least, a 3D representation though usually includes a plan and elevation "view". If there is no plan definition most systems use the plan projection of the 3D view. When assembled in the building, issues around access, fitting, buildability, joining and clashing can be determined from the different views. Some of these processes are automated. Clash

---

6   "BIM handbook", Eastman et al, Wiley, 2011, page 16

7   "Modelling Multiple Views of Buildings", Nederveen and Tolman, Automation in Construction, 1992

8   "Building Information Modelling - The Great Debate", http://www.laiserin.com/features/bim

9   REfLEX Review June 1995, Page 11

detection will check to see if parts intersect. For instance, does any duct work run through beams? Design issues can be identified and resolved.

In order to design a complex joint, it may be crucial to have complete three-dimensional information about each part and be able to move, adjust and view those parts in a comprehensive and changeable way. In order to determine if that part can be fitted while the building is being constructed, a 3D view of the partially constructed building may be required from the associated construction assembly.

All design information can be extracted from this BIM model. This includes, but is not limited to, conventional working drawings, 3D views, schedules, costings and quantities. All changes are represented fully and automatically. The BIM model can be extended in terms of data stored or referenced to ensure that it can always produce the coordinated information.

It may be necessary to have access space around a given object in the building, sufficient to allow maintenance or access. This can be built into the virtual object representing that part so that it will flag clashes.

COORDINATED INFORMATION

As stated earlier, any data storage system that has duplicated information can lead to an "information clash". This happens when one figure is changed and another is not. For instance, if your address occurs in different databases there is no guarantee they will be the same in each database. If they are different, which is correct? In a BIM model, the second occurrence of the address is replaced by a pointer to the first.

In a building drawing or model, a dimension that is duplicated may at some stage be changed in one place and not the other. For instance a staircase leading up to a slab floor might store the depth of the slab stated either implicitly or explicitly. If the slab depth is changed then the staircase will have an incorrect depth and this will probably be revealed on the construction site. This problem is solved within the BIM model.

Sometimes this will mean going to an external database of values, perhaps of dimensions, construction dates, costs or materials, or to an adjacent component in the model to retrieve this information every time the design is redrawn. In the staircase example in the previous paragraph, the staircase must ask the slab each time it is drawn or accessed how deep the slab is, in case it has changed.

Parametric components themselves have rules, knowledge and detail built in that enable this type of communication and allows sophisticated rule checking. As an example of rule checking, our staircase could have some reinforcement detail associated with it. A sophisticated staircase object will look at the floor area it is servicing and determine its minimum dimensions by checking against the appropriate standards and adjust the reinforcement accordingly.

The use of parametrics is very much a part of BIM. Variable design, self checking connections and passing information can only be achieved with parametric objects. The passing is done invisibly. A system without sophisticated parametrics cannot be described as BIM and will fall at the first hurdle of changing data.

Much or all of this linking is hidden by the BIM system. For instance Revit, automatically joins the walls and doors and takes care of the links automatically. Figure 19 shows coordination of the different drawings extracted from the single model. Note that there is a mixture of 2D and 3D representations and that many of the shapes are parametric. Non-graphic information was also generated from the single model. Figure 8 illustrates a simple door

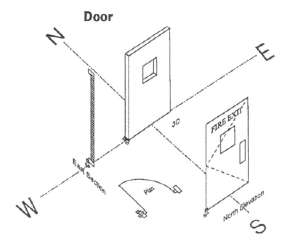

Figure 4. *Default Single Door with different type of views*

schedule, effectively a different view again. This was placed on the drawing or sheet and regenerated automatically each time the drawing was viewed or plotted.

3D MODEL

People speak about BIM and 3D in the same breath as if they are one and the same. Rendering images is an important part as it provides immediate visual feedback as to the state of the model and information within the model. Rendering might also include mapping data, climate data, satellite imagery, photographs, augmented reality (AR) information, radar mapping of the underground, vegetation maps, and landscaping, aerial images and laser scans of infrastructure. A BIM contains a 3D model but much more besides.

The BIM model provides a framework on which to hang this type of information. It helps coordinate and bring the complete picture to the design environment. Figure 120 (Heathrow Express) and Figure 137 (Ground Modelling) show ground data displayed inside the BIM model. In this case the Author wrote parametric components to read the data points into the model and display the interpolated surface as either contours or a triangulated surface within the BIM.

BIM CONCEPTS

Each member of the design team looks at the building in different ways and the drawings produced vary in style, content and intent. As the design concept is developed, the single idea of the building at inception is expressed in an increasing number of drawings and written documents.

Assembling a BIM model may be loosely compared to working with Lego, or perhaps Minecraft, but using more realistic parts. As stated earlier, the evolving design is represented at all times by the developing model and when retrieving information or visualizing the design in any way, the latest design is all that is available, ensuring up to date information.

Construction drawings consist of different types of graphical information that represent the components or elements of the building. Language and functionality typically dictate the "extent" of these components.

Figure 4 shows the door and its different definitions or "views". The 3D view of the door would have surface properties included so that it could rendered. Figures 7 and 9 show the different systems placing the door. The door also cuts a hole in the 3D view of the wall and has closure on the cavity details in both systems.

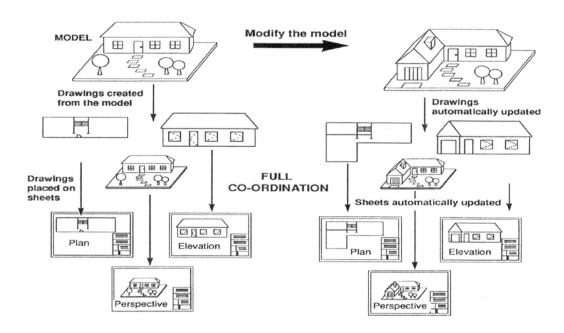

Figure 5. *Different Drawings affected by change of design © SONATA Systems 1988*

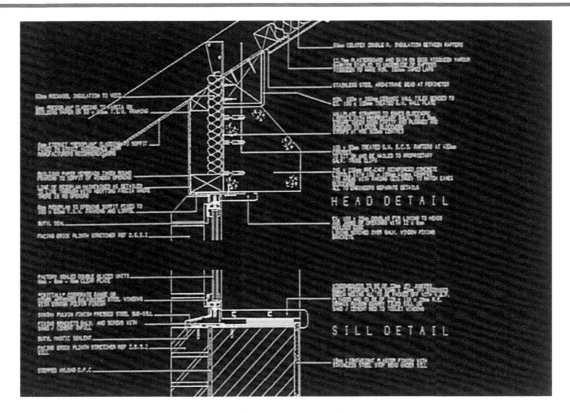

Figure 6. *Screenshot SONATA Elevation Detail (automatic section)*

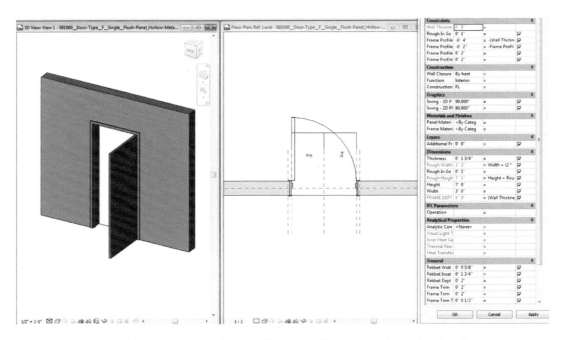

Figure 7. *Revit Door with parameters similar to REfLEX in next figure. Note the similarities of cavity closure and hole in the wall*

Figure 8. REfLEX User defined Parametric Door in cavity wall 1996 © Peter Dew and Associates

PROJECT TITLE: TOWER POINT, OFFICE DEVELOPMENT
JOB NUMBER: J001/91
CAD PROJECT NAME   TOWERP
TAKE OFF BY: C. McCarthy

DOOR REF NUMBER	OPENING SIZE (mm)			HUNG	LEAF WIDTH		FRAME DESCRIPTION
	Width	Height	Thckns		LEFT	RIGHT	
DB1/1	900.0	2100.0	100.0	left	800.0		Single leaf, single swing door
DB1/2	1200.0	2100.0	100.0		900.0	200.0	One & half leaf, single swing door
DB1/3	1800.0	2100.0	100.0		850.0	850.0	Double leaf, single swing door
DG/20	1000.0	2000.0	150.0	left	900.0		Single leaf, single swing door
DG/22	2100.0	2100.0	100.0		1000.0	1000.0	Double leaf, single swing door
DG/24	1500.0	2050.0	150.0		1050.0	350.0	One & half leaf, single swing door

Figure 9. Simple SONATA Door schedule 1990 placed on the drawing sheet, updated every reference

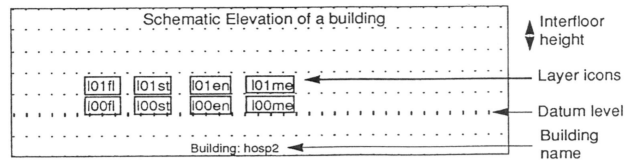

Figure 10. *Layers placed at different datum in the building © SONATA Systems 1988*

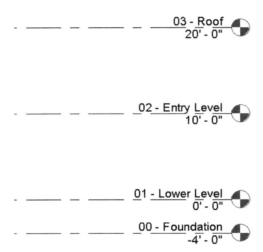

Figure 11. *Revit data objects defining the levels, grids and extent of information © Autodesk 2016*

End and cavity details are generated automatically depending upon wall/door/window details and dimensions. One of the issues with parametric systems is the amount of detail that needs to be generated when you place an item. For instance the door has a lot of questions about supplier, iron mongery and fire rating that you may not want to enter at the moment of placing the door for the time. The amount or quality of information in a BIM object is called the Level of Development in the US and Level of Detail in the UK.

The BIM system knows which of the views to access for the appropriate drawing. In a plan you tend to get plan views (but not always) and in a 3D view, always a perspective view. It is a little more complex than might be evident, for example what if you have an object rotated onto its side where the plan view is

not representative of the object. This is all handled automatically by the system. The Author spent some considerable time first implementing what views of an object to see at different rotations.

Components can also look different at different scales in terms of content. This means they may represent themselves differently depending upon the scale of the drawing or the closeness of the viewer. A 1:10 view will be different to a 1:100 in terms of hatching and detail. A shelf of products will look different close up than across the store.

Similarly it should be possible to vary the section of an object depending where the section line is cut and the depth of section is set. (SONATA and REfLEX did). The BIM object has access to various systems variables such as the scale or where the section was being cut and so can represent itself correctly.

An important part of BIM is that the component parts can self-check and self-design. These benefits allow details, even complete parts of buildings, to be designed automatically. Figure 115 and Figure 117 show the core of a building modelled as a complex parametric and a complete stadium as a single parametric. Building Code checking and local consistency can be built into such objects. In order to build such information into a component, one must go beyond the graphical construction and use either a language or an external system to graphically generate the complex object. Depending upon requirement a number of different systems exist to assist with this, including Rhino, Grasshopper and others.

Another advantage of parametric objects is that they can store non-graphical information that needs to be used for purposes other than representation. For instance, delivery dates, suppliers, costs and so on can all be built into the parametric data. This leads to thermal calculations, stress and strain, fluid flow rates, construction management.

Figures 146, 151 and Figure 157 show objects that have construction dates associated with them. Tools are provided to view the model at different dates, in this case with a slider bar. Similarly Figure 167 shows objects and processes (also objects) linked to a Gantt chart. Changes to the dates, times or sequences changes the Gantt chart and vice-versa.

Building components are usually assembled onto layers in the virtual building. Each layer is effectively a work-plane set at a particular datum where particular types of components can be assembled. Different disciplines tend to use different layers traditionally using layer locking to protect changes from outside disciplines.

Layers are assembled at different heights in model. Figure 10 shows the original use in SONATA and Figure 11 shows the Revit use of datum objects.

As we now know, all drawings are produced directly from the BIM model. The drawing definition itself consists of references

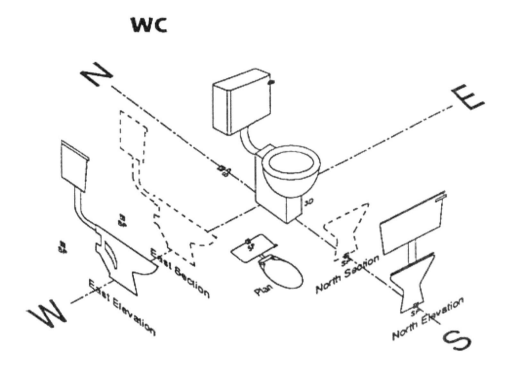

Figure 12. *Views of a Loo, Principles and Applications © SONATA Systems 1988*

to the areas and floors of the building being drawn, the type of projections, and a selection of information and layers of what is to appear. Sheets of drawings, in the traditional sheet definition, are assembled graphically. Figure 19 shows a sheet of SONATA drawings. Note the mixed symbology and 3D in the plans and sections, and the pure 3D in the perspective view. 2D objects were permitted and given a tag FACE VIEWER. This means that something like a tree or a person does not have to be modelled in full but can be done as a cutout. The disadvantage is that in a movie of the scene that includes the figure, the figure can be seen to rotate.

In some systems changes are "bi-directional" where you can change a dimension on a drawing this will affect the model. Revit has this capability. You can change the drawing to change the model.

One does not save old drawings (except for reference), drawings and schedules are generated afresh every time they are needed. If you make a change to the building model, then every single reference (including drawings) to the model taken after that moment reflects this change.

One of the features that is treated differently in the various systems is how to generate historical drawings. Revit stores this information within the building model. The issue with this is that the model can become very large making the system unusable in some cases. [PS There are fixes involving saving with a new date.]

Generating schedules is a very similar process to generating drawings. An area or view of the building is used to define the

limits of the schedule, potentially the whole building. Relevant information can be collected from the information part of the model and displayed separately from the drawings, or perhaps pasted onto drawing sheets. Again the schedule is never saved except for reference and is regenerated when required. Revit has specialist tools for generating BoQ to the various standards.

INFORMATION PASSING BETWEEN BIM ELEMENTS

Considering the idea of information passing further, in the earlier example, a window frame in a wall may have been designed to have the frame as thick as the wall. If the wall is changed in thickness and the window has no reference to the wall, then the window, as we know will be out of date. It may have the wrong frame. The wall needs to know there is a window of some size to make the correct sized hole, adjust the energy calculations, and change the bill of quantities, and the window may need to know something about the wall, is the wall external perhaps?

When viewing the image, the window must make a "hole" in the wall so one can see through the wall and that wall quantities are correct. This is not a trivial exercise, as the hole must match the size and shape of the window, door, or indeed whatever is being inserted into the wall. This is done by a "Boolean Operation" on the different 3D shapes. With this, one can add and subtract shapes as necessary.

Joining multiple cavity walls at a single point in plan, elevation and 3D must also be dealt with. The walls, the hatching and the cores must match and stop at the appropriate place. Obviously the system must deal with hatching and cavities and any 3D implications. SONATA, REfLEX and Revit all do this join.

Figure 13. *SONATA Image Land Securities Grand Building Image by Mark Edwards of CADAIM © CADAIM*

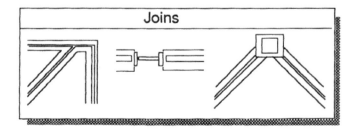

Figure 14. *The "Join" menu Icons from the SONATA Manual 1988 © SONATA Software Systems Inc*

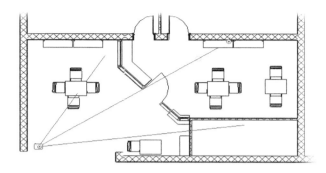

Figure 15. *Revit Camera tool © Autodesk 2016*

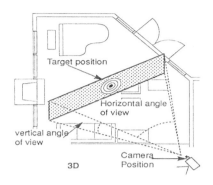

Figure 16. *SONATA Camera tool (annotated) 1988*

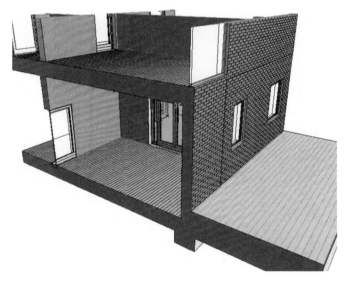

Figure 17. *Archicad Section planes © Graphisoft 2009*

Figure 18. *SONATA Image showing variable Section Planes 1992*

Figure 14 shows a menu for explicitly joining the walls and or windows in different ways.

The concept of information can be used to gain additional design and self solving design capabilities. In implementing this "information passing" functionality in SONATA in 1985, the Author extended it to form networks of components linked together. This was done by exchanging networks of lines for the appropriate components. For instance when making a duct network, two lines forming a "T" was replaced by three straight pieces and a T piece. So the duct network could be set out with the pieces, straights, bends, T pieces, AHU and so on all inserted automatically. A "physics" view was also added to each component. This view balanced the changing properties at the entry/exit points on the

body. In the case of the duct network it was volume, temperature and pressure and in the case of steel members it was loading.

The "Solve" button caused this network to "relax" producing a solution to the network. The "Relaxation Method", allowed the ducts to iterate to their correct sizes given the standards and the physics. Figure 172 shows a snapshot of the different pieces in a duct network during the iterative solution process. Using the flow rate into each room, the flows and sizing are determined across the whole network. The sizing is determined for each piece according to the tables in the standards and Air Handling Unit (AHU) capacity is determined. Changing the air change rate in a room would have an automatic knock on effect back to the

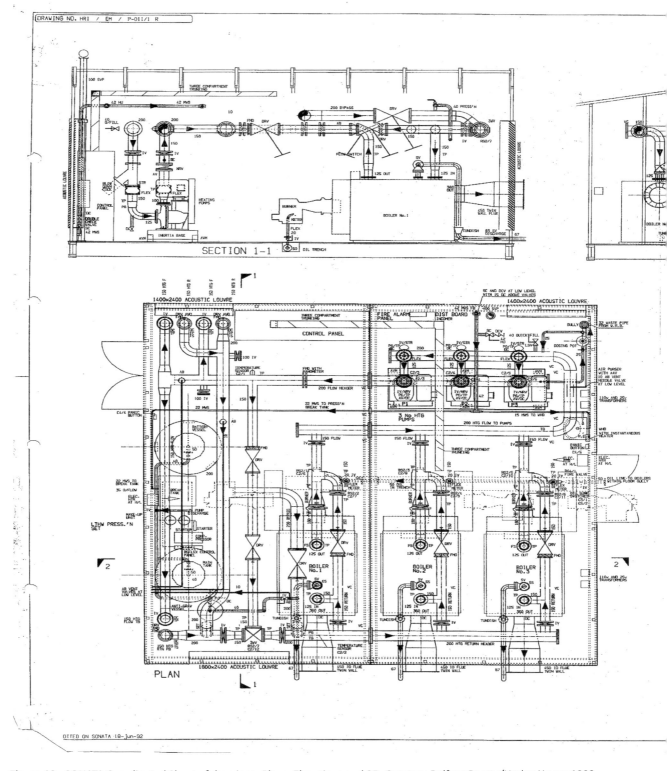

Figure 19. *SONATA Coordinated Sheet of drawings; Plans, Elevations and 3D, Courtesy Balfour Beatty/Haden Young 1992*

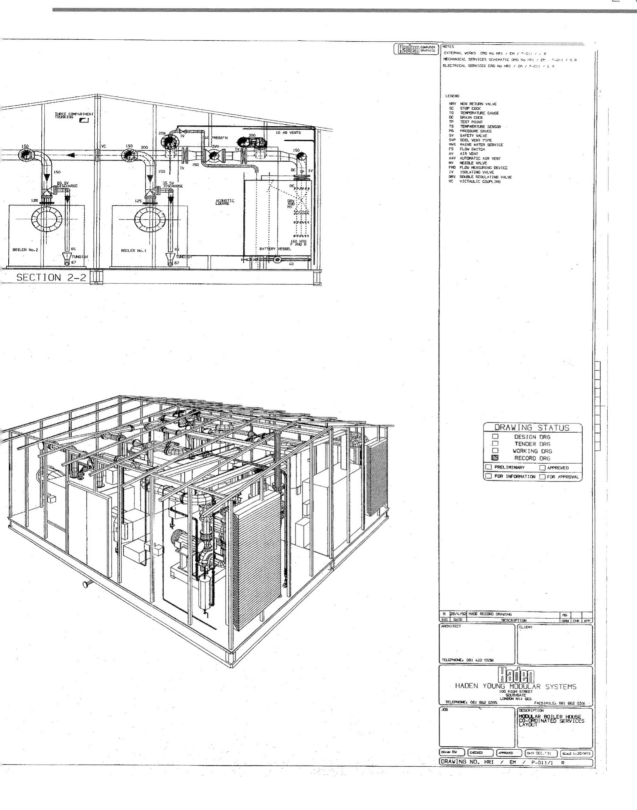

AHU. Slow machine speeds in the 1980s restricted the use of this capability.

Further information on this technique can be found in Appendix 1. Networks could also be moved and treated as an entity. It is also interesting in that each object behaves how it should without reference to other objects in the network. Zampi referred to this functionality in his book "Virtual Architecture".[10]

FAMILIES

BIM systems can group components together into families. Family members have a related graphical representation and tend to belong to the same functional groups.

Some systems use user defined groups, allowing a graphical table of components to be shown as a menu. Groups could also be selected using keywords or categories.

Revit extends the concept of families to cause automatic linking between appropriate families. For instance if a member of a wall family has a window family sitting in or near it sets points and effectively automatically joins them. In defining a family, Revit requires that the variables or parameters associated with the component have common names and meanings. Examples of a Revit family would be a wall where different wall types with the same basic parameters are grouped together.

---

10   Virtual Architecture, Zampi and Morgan, Batsford Books 1995

Figure 20. *SONATA teamNet sharing database concurrently © SONATA Systems 1988*

Although these families serve different purposes and are composed of different materials, they have a related use. Each type in the family has a related graphical representation and an identical set of parameters, called the family type parameters.

Revit has three different types of families, which seems an unnecessary complication:
- place families that are available only in the project they are created and are dependent on the model geometry
- system families that are hardwired and you can only change the parameters
- component families that have been created in the graphical editor

IMAGE RENDERING

Visualizing the building model is an important part of BIM systems and is most often used. A rendering program takes a perspective (usually) view of a computer model and generates a colour image of that view.

Today, hardware based graphics boards can render the most complex models fast enough to appear to be smooth movement. As an example Figure 165 is an example of VR from the 1990s and Figure 185 and Figure 197 taken from a VR system built by the Author more recently. The quality of the best images is reaching a level where it is difficult to tell what is real and what is not.

Out of interest, there were no rendering algorithms available either in software or hardware in the 1970s or early eighties. In 1976 the Author had written a rendering algorithm (see Appendix 1) which he had wanted to commercialize. He approached Evans and Sutherland with this technology and when this failed to attract their attention, he decided to use this knowledge to build a new modelling system for engineering and architecture. The SONATA images use this algorithm (except figures by BAM Construct in 2010).

In order to achieve a colour image of a building, the computer program or graphics board, takes the 3D data, works out what is visible, how the light falls on the surface, looks at the surface properties including textures and shows the appropriate colours on the screen, pixel by pixel. This means that each visible face, contributes with facet orientation, facet colour, facet texture at the point, reflectivity, shadows from various light sources, light sources, light colours, light positions, fog colour, density and distance to the viewer and spot light colour, orientation and intensity. The mathematics and algorithms are complex and are well hidden from the end user. Anyone can tell the success or not of such algorithms simply by glancing at the end product.

Figure 21. *SONATA Hidden Line St Ethelburga's Church, Bishopsgate © Blee Ettwein Bridges Partnership 1995*

Rendering essentially means taking a mathematical "photo" of the model of the building. Making an attractive rendered image requires careful consideration of the building itself, the surface colours and textures and the lighting. Rendered images can be harsh and many a beautiful building has been rendered badly to produce an ugly image.

Hidden line images are also an important part of the output for any BIM system. Detail of complex constructions can be included on the 2D drawing sets and removing the use of colour brings it closer to a working drawing. Usually hidden line views are less emotive than the colour, one sees the detail rather than the image as a whole. The beautiful image of St Ethelburga's Church shown in Figure 21 is the contradiction to this rule. Some of the complex services drawings use hidden line. If you look carefully Figure

180 has duct obscuring or hiding duct at different parts of the drawing.

Methods of generating a hidden line view are similar to rendering in terms of specification of the view and information. The actual algorithms are very different and in some ways more complex. The Author produced a movie of a real street scene in 1976 in hidden line. This was one of the first architectural projects to use a computer generated movie in order to win planning approval. A shaded image was not used as there was too much data for what was then one of the largest machines available (CDC Cyber 76).

So far we have seen a number of technical capabilities that shape the usage of the BIM systems. These characteristics allow, encourage and in some ways enforce many of the changes in procedures found in BIM usage in practice.

But BIM is more than just a change in technology. It involves change in procedures, work processes and significant changes in the legal practices.

COLLABORATION & SHARING THE MODEL CONCURRENTLY

Using a shared model, engineers, designers, owners and contractors can work through the design and achieve trade-offs and improvements in design and costs as well as buildability. Owners can be engaged throughout the entire process, and contractors can use the model to create an informed bid. All parties can consider the detail of the structure, and can discuss and walk-through how to construct the design.

The model can also be used to communicate project details to the public. Rendered presentations of the structure provide a compelling and intuitive understanding to non-technical people.

In some ways BIM technology enforces collaboration between the different stakeholders. It is a playpen where everyone must play with their work on display. It brings down the firewalls and ensures an openness, removing distrust and provides a forum for Active Design.

But it is not a free-for-all when using the model. There is always an audit trail of changes and there are access restrictions between the different layers, allowing for, say, only the HVAC engineers to change the duct work.
This collaboration in sharing the model brings a number of benefits.

- The shared model gives managers and clients a design overview.
- It enables changes to be made quickly without manually passing information backward and forwards. These documents can quickly go out of date leading to issues in the design.
- It improves collaboration between the different parties.
- It enables higher level engineering such as sustainability checks and validation to be performed over the shared model.
- It provides an open environment for the different parties engendering cooperation and good will.

A shared model has other benefits after the design phase and as the construction progresses.

The coordination of ordering, bringing materials on-site, and the actual construction of specific parts can be managed and controlled through the shared model. Different disciplines control the these aspects (as opposed to design) with a view to making sure the construction is carried out efficiently.

Once constructed the model can then again be shared. An engineer on site may use an augmented reality view into the model, the emergency services might have a view of the fire systems, the facilities manager would have a view of equipment location, and management would have an overall view of the site.

Sharing of the model is usually done through the cloud, an extra-net or a local server. The SONATA model was shared to different sites as early as 1988. See Figure 162 and Figure 20.

COMMON DATA ENVIRONMENT

The model and associated information goes into a shared space called the Common Data Environment (CDE). The data that resides in the CDE includes, the model, graphical and non-graphical assets. This collection of digital files may be spread around the globe although it appears to be in the same virtual space. BIM components have the capability of referring to non local files across the Internet.

Sharing the model and the CDE ensures a continuous system integration across all parties involved. It also ensures the sub-contractors and suppliers are coordinated in terms of materials. Further in the book it will be shown that the coordination in a temporal or time sense allows for them to be coordinated in terms of "when", and that deliveries can be scheduled around works in time and space, understood through the sharing of the model.

DESIGN AND ENGINEERING

The parametric nature of BIM brings many benefits particularly to the design and engineering phase.

- The components can have rules and regulations built in and they can refer to external tables ensuring regulatory compliance.
- Components can refer to external files for manufacture and supply information ensuring up-to-date information associated with the model.
- Components can be easily imported from manufacturers.
- Perhaps obviously, component numbers are limited by having variable parameters.
- Clash checking can be integral to components.
- Constructibility can be determined by parameter and examining conditions around the installation area.
- Engineering detail such as reinforcement can be determined from within the components themselves.
- Optimizing design has been used to ensure the best possible component shape and size.
- Components can adapt themselves to the changing environment of the surrounds.
- Integrate work-flows easing design error.
- Easy evaluation of variable designs.
- Optimized spatial design can be achieved by allowing automation of arrangements and components.
- Integrating different disciplines automatically, such as road, tunnel, and surface design with radar, laser and road layout schemes.
- GIS data is available for integration with parametric elements for large projects.
- Iterative design through linked networks of components.

CONSTRUCTION SEQUENCING

Assigning date information to components in a building and providing tools to view the building based on those dates creates a time base associated with the building. Associating manufacture date, order date, delivery date, transport to site date, construction start date, finish date and so on with the components means the building or the order and construction process can be viewed as part of the model, tying together time and space. Some of this information may be referenced on the manufacturer's site so not all is carried in the component model. If a component is changed in the model or even a parameter, the delivery time or construction time might change. This would be automatically translated into associated Gantt charts and building construction time. Chapter 11 on Case Studies in Construction Management shows a wide use of this type of information.

The construction phase of a structure can be modeled seamlessly in BIM; it enables the simulation and sequencing of activities in time and space. From this simulation and sequencing come scenarios for procurement, spatial, man power, resource and sequencing the work. As shown in later chapters, principles of "lean construction" taken from the car industry and Just In Time procurement can be applied. BIM also encourages more collaborative forms of project delivery such as integrated project delivery and design build. Further savings can be achieved with these proven techniques. Management of sub contractors and suppliers can be enhanced using BIM.

Tendering can be made more transparent, accurate and efficient, because of the large amounts of information available in the BIM model to the potential bidders. In addition, design changes as they happen can be furnished to all parties, ensuring accuracy and reducing requests for information.

Live Gantt charts, manpower requirements, automatic ordering and visualization of delivery can be achieved during the actual construction process. This ensures correct sequencing of deliveries and assembly of materials and parts into the structure. Another capability in simulating the construction sequencing is that of structural integrity, as it is constructed, the plant can be tested to ensure strength, curing and so on. BIM can also assimilate other technologies, such as laser and radar readings.

Live monitoring of the construction site can be achieved through sensors bringing information into the model. This was achieved on-site at the Heathrow Express coffer dam in 1996. Sensors measuring movement in the coffer dam were linked to the model.

BIM works well in prefabrication. Components representing the prefabricated pieces exactly can be maneuvered and manipulated.

OPERATIONS

BIM facilitates operations and maintenance of a building by providing a virtual, potentially working model of it. The model has information stored from the design and construction phases. It may also have time laced information embedded into components in the model, such as a scheduled maintenance procedure. Integration of the model into Facility Management and Asset Management ensures significant operational savings. BIM can ensure that information is not lost in the handover and commissioning. This can be prepared virtually before the actual handover, with documentation, manuals, maintenance procedures being handed over with the build model. Linking RFID devices or bar codes with the model can help keep track of equipment and materials.

BIM can be used as an enhanced reality tool to show hidden pipes, equipment or even instructions, while maintaining equipment in the completed building. This is discussed further in Chapter 11, the Future.

Embedded sensors, fire systems, lift operations, CCTV and so on can all be focused back into the as built model. This all enhances the operations and management procedures. Refits and renovations can be greatly simplified by using an as-built BIM model. See Chapters 9 and 14 on Retail Case Studies and Retail Information Modelling respectively.

In this chapter we have seen many of the component parts of BIM. Some of these parts are technical and others are how the work is done around the technical. BIM is the summation of the technical and human parts.

In the next chapter we look at the history of computerized Building Design, leading to how the first BIM system was developed.

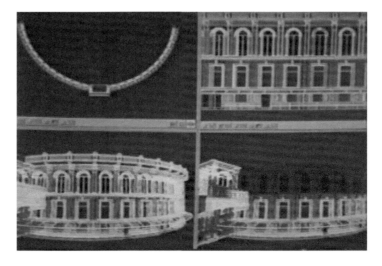

Figure 22. *REfLEX Screenshot of the Royal Albert Hall*
© *Taylor Woodrow 1995*

Taylor Woodrow and BAA at Gatwick North Terminal 1995

Perhaps the first use of BIM with true collaboration between all stakeholders on large scale, was achieved in the building of the Gatwick Airport North Terminal Facility. This project was innovative in several respects; it was used in initial presentations to win the work, it used a BIM model to represent all design aspects of the structure and that became part of the process adopted by the owner BAA. At any early stage, formal evaluation of the BIM model was undertaken by the primary stakeholders, the engineering and maintenance staff.

Baggage handlers, airlines, Her Majesty's Customs and Excise and the Department of Transport Police all had input into the model at this point. Workshops were held using the BIM model and a design was settled upon. In one of these meetings customs authorities identified inadequacies in the passenger flow layout with potential security breaches allowing for changes to be made to the model.

After the workshops, a meeting with all stakeholders was held and a legal agreement was signed to ensure commitment and zero change orders.

BAA's experience was such that this process enabled proper project control, allowing a facility to be built with no change orders. Also noted by BAA was that the concurrent engineering involved suppliers as part of the integrated teams and their commitment to pre-planning of fabrication and construction. In addition the model was used successfully for public relations.

Further details of this project can be seen in Chapter 11 BIM and Construction Management.

Background to BIM

3

Computers and software have been developed to solve engineering and design issues since the late 1960s, but perhaps the first use of graphics was the display of radio signals generated by lightning strikes. This was done on an oscilloscope at the Meteorogical Research Station, Aldershot UK, by Robert Watson-Watt in 1925.

Arguably the first computer, built by Konrad Zuse in 1941, was used for aerodynamics calculations by the Axis powers. Britain developed radar during WW2, and the first truly programmable computer became available soon after, in 1948. It was from this time that basic screens started to appear, Whirlwind perhaps being the first to attach a Cathode Ray Tube (CRT) for air-defence.

Ivan Sutherland of Salt Lake City, Utah, U.S. is considered the inventor of interactive graphics with the first "Robot Draftsman". This he called Sketchpad[1], enabled the user to set, draw and interact on a CRT screen with a light pen, the precursor to the modern graphics interface. Sutherland went on to form the first computer graphics company, Evans and Sutherland in 1968.

From the mid-1960s, various Computer Aided Drafting (CAD) systems started to appear. Only the largest companies could afford computers capable of performing the calculations required fast enough, typically for the Automobile, Aeronautics, Defence and Electronics industries.

In the 1970s, development of Computer Aided Design (as opposed to Drafting) took two separate routes; one towards general-purpose geometric drafting systems, and another to more specific building type modelling. Meanwhile, large engineering corporations created drafting systems based around 3D shapes,

dealing with complex curves, with output direct to Computer Numerical Control (CNC) machines.

While 3D engineering drafting model approaches were mostly developed in the U.S., building-specific CAD research was mostly done in the UK. BIM systems were certainly their precursors intended to measure the performance of both existing and future projects, compute their costs and research generic design issues. Quoting Prof Sir Leslie Martin when he set up the School of Architecture (SoA) Research Centre for Land Use and Built Form Studies (CLUBFS) in 1967, with the aim of modelling architecture, planning and building, *"We become aware of another way of looking at a design problem through which we can consider more effectively and rigorously the ranges of choice that are open to us."*[2]

While it seemed to many that they had hoped to computerise design itself, that wasn't the case, as a recent essay by one of the founding members, Philip Steadman[3], sets the record straight in reply to Two Cambridges: Models, Methods, Systems and Expertise by Mary Louise Lobsinger[4], and Sean Keller's Fenland Tech: Architectural Science in Post-war Cambridge[5], clarifying that they had only intended to computerise an understanding of building, especially environmental performance and building costs.

---

1   "Sketchpad; A man machine graphic communication system", PhD Thesis, Sutherland, 1963

2   "Geometry of the Environment", Lionel March and Philip Steadman, RIBA Press, 1971

3   "Research in architecture and urban studies at Cambridge in the 1960s and 1970s: what really happened", Routledge, Taylor & Francis Group, March 2016

4   "A Second Modernism", MIT, Architecture, Edited by Arinfam Dutta, MIT Press

5   "Fenland Tech: Architectural Science in Post-war Cambridge", MIT Press Journals, Grey Room, #23 Spring 2006

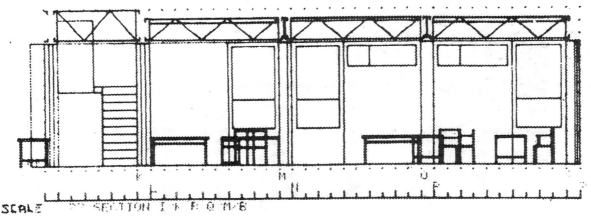

Figure 23. *System BDS symbolic elevation view from "Computer Applications in Architecture", John Gero, Applied Science Publishers 1976 © Elsevier*

The research into Building Design was largely government funded, either through the National Health or Universities. In the case of ARC discussed below, the first university spin-off company was established. Various systems were developed around the concept of a building model with a support suite of applications, where each system defined a single coherent view of the building around which applications were clustered. Most of these systems were specifically for building design and drafting, which have been largely excluded from this chapter as being general-purpose systems of the time, and in turn superseded. Eastman's book Building Product Models[6] gives a comprehensive view of most

of the early building modelling systems: here, the aim is to find common threads with today's BIM.

Each of the systems discussed had a common vision: to assist the design and construction of buildings by first constructing a virtual building. Most systems differ significantly from each other, although one or two are clearly dependent on the others. The most significant systems of the time were developed by Applied Research of Cambridge (ARC) for hospitals, CEDAR from the DoE for post-offices and SSHA from Edinburgh for Housing.

BDS BY APPLIED RESEARCH OF CAMBRIDGE

ARC, established in 1969, was the first university high-tech spin-off company, and went on to develop OXSYS[7] specifically for the Oxford Regional Health Authority in 1971. Later called BDS it was described at the time as *"it could be used to produce a* [building] *definition in 2½ dimensions. From this definition drawings could be made together with a wide range of supplementary results".[8]* It was based around a set of predefined hospital building elements and was used to define buildings defined by the Oxford Method industrialized system. ARC introduced various new ideas involving hierarchy of elements and parametric properties, with a predefined set of components including structural and cladding elements, as well as interior fixtures and fittings.

---

6   Building Product models: computer environments supporting design and construction, Charles M Eastman CRC Press 1999

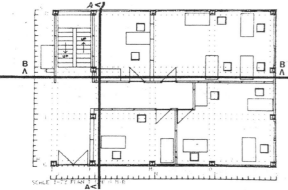

Figure 24. *System OXSYS plan view from "Computer Applications in Architecture", John Gero, Applied Science Publishers 1976 © Elsevier*

---

7   "Computer Applications in Architecture "- John Gero, Applied Science Publishers, 1977, Page 343

8   "Computer Methods for Architects ", R.A Reynolds, Butterworth & Co., 1980

34:1:4   FACES
HAND BASIN TEST COMPONENT

Figure 25. System BDS Concrete views of a single component from "Computer Applications in Architecture", John Gero, Applied Science Publishers 1976 © Elsevier

EXAM/TESTING

FUNCTIONAL GROUP OF SPACES	NUMBER OF ROOMS	REALISTIC DIMENSIONS			PHASES (UP TO THREE)	SPECIAL CHARACTERISTICS							ROOM USE CODES	FLOOR, WALL, CEILING MATL. HVAC, ELECTRIC, PLUMBING			
	NUM	LEN	WID	CST		P H S	F R M	E X P	C L G	F L N	C O V	E N	FNCTN	QULTY	UNIT AREA	ITEM AREA	COST
020201 ELECTROENCEPH	1	10	11	98		1	2	0	9	0	2	8	EMS11	6M905	110	110	10780.
020202 EEG EXAM TEST	1	10	11	86		1	2	0	9	2	2	8	EMS12	1N344	110	110	9460.
020203 GENERAL PURPOSE	1	10	11	123		1	0	1	0	2	1	8	WHS42	1A374	110	110	13530.
020204 ELECTROCARDIOG	2	10	11	98		1	1	0	9	2	2	8	EMS13	1A365	110	220	21560.
020205 INHILATN THRPY	1	10	11	91		1	1	0	9	2	2	8	EMS16	1A024	110	110	10010.
020206 EFFORT TOLERNC	1	10	18	61		1	1	0	9	2	2	8	EMM21	1A004	180	180	10980.
020207 PULMON FUNCTN	1	10	11	85		1	3	0	9	2	1	8	EMS12	1M315	110	110	9350.
020208 DENTAL EXAM	1	10	11	92		1	1	0	9	2	2	3	EMS01	1N040	110	110	10120.
020209 ENT EXAM TRTMT	1	10	11	98		1	1	0	9	2	2	3	EMS01	1N040	110	110	10780.
020210 EEG OFFICE	1	12	23	51		1	1	0	9	0	1	8	OTS14	6A304	276	276	14076.
020211 EEG READING	1	5	9	0		0	0	0	0	0	0	0			45	45	0.
020212 BLOOD GAS	1	7	11	0		0	0	0	0	0	0	0			77	77	0.
020213 WORKROOM	1	10	88	51		1	1	0	9	2	2	8	WHS42	1A014	880	880	44880.
CIRCULATION	50%			45											1224.		55080.

DEPARTMENT 0202 EXAM/TESTING

	PHASE ZERO		PHASE ONE		PHASE TWO		ALL PHASES	
	AREA	COST	AREA	COST	AREA	COST	AREA	COST
NET	122.	0.	2326.	165526.	0.	0.	2448.	165526.
CIRC	61.	2745.	1163.	52335.	0.	0.	1224.	55080.
TOTL	183.	2745.	3489.	217861.	0.	0.	3672.	220606.

Figure 26. ARK-2 COMPROGRAPH Hospital rooms, space and cost usage from "Computer Applications in Architecture", John Gero, Applied Science Publishers 1976 © Elsevier

BDS also defined a building by having the user divide it into self-contained zones. These zones, categorized into different types, were defined by their boundaries. Properties were assigned to zones. Starting with partition walls, the overall layout of the structure is determined. Components could only be placed in orthogonal orientations and the user had to be aware of the data structures and the organization of the data.

The predefined elements were structured by family type into a hierarchy, with lowest points on the hierarchy being represented by 2½D components, typically being several plans, elevations and sections. This idea, first used in BDS, has become pervasive in modern BIM, called box-geometries or "Concrete components". Geometrically, each component was defined by its projections onto the surfaces of an enclosing box. See Figure 25. Elaborate

```
      -RADST-UPTAK-SCAN -DIRCT-PCHGE-SBWTG-CAMRA-ULTRA-RECEP-STLT -
LAB   -  3     2     3     3     5     3     3     3     5     3
      -RADST-UPTAK-SCAN -DIRCT-PCHGE-SBWTG-CAMRA-ULTRA-RECEP-
STLT  -  3     3     3     2     2     3     3     3     1
      -RADST-UPTAK-SCAN -DIRCT-PCHGE-SBWTG-CAMRA-ULTRA-
RECEP -  1     3     3     2     3     1     3     3
      -RADST-UPTAK-SCAN -DIRCT-PCHGE-SBWTG-CAMRA-
ULTRA-   5     3     1     3     2     4     3
      -RADST-UPTAK-SCAN -DIRCT-PCHGE-SBWTG-
CAMRA-   3     1     3     3     2     4
      -RADST-UPTAK-SCAN -DIRCT-PCHGE-
SBWTG-   3     3     3     3
      -RADST-UPTAK-SCAN -DIRCT-
PCHGE-   3     3     3
      -RADST-UPTAK-SCAN -
DIRCT-   2     4     4
      -RADST-UPTAK-
SCAN  -  3     3
      -RADST-
UPTAK-   3
```

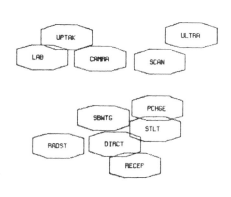
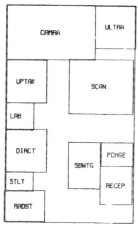

Figure 27. *COMPROPLAN Bubble diagram of room relationship & Manipulated block diagram from "Computer Applications in Architecture", John Gero, Applied Science Publishers 1976 © Elsevier*

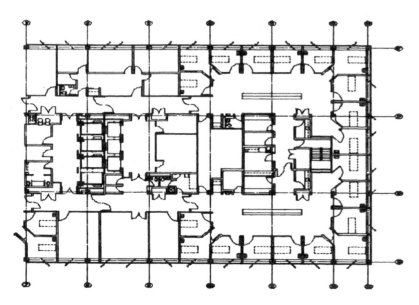

Figure 28. *ARK-2 COMPROSPACE final plan layout from "Computer Applications in Architecture", John Gero, Applied Science Publishers 1976 © Elsevier*

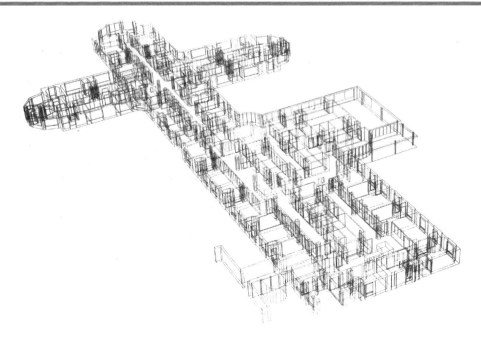

Figure 29. *ARK-2 COMPROVIEW Interior Perspective View from "Computer Applications in Architecture", John Gero, Applied Science Publishers 1976 © Elsevier*

rules determined their behaviour as they were hard-coded into the program, from which ARC updated and refined the elements. This collection of elements, called a CODEX, was extensive and included many different types of elements.

OXSYS itself was divided into three parts:

- Basic Operating System (BOS) which dealt with low systems level databases, graphics and user interaction, none of which were building specific.
- Building Design System (BDS) contained a range of programs for general design tasks.
- Detailed Design System (DDS) dealt with the automatic component location and selection using the "Oxford Method" of design related to the particular project under consideration.

OXYSYS was based around the idea of symbolic views for each object. Unfortunately, neither the original nor the more commercial BDS supported user-defined parts or components, so it was withdrawn in 1986 *"when full 3D and solid Modelling geometry became integrated into production systems."* Ed Hoskins the CEO of ARC had concluded that it was no longer viable as *"an established practice will not reorganize itself around a new, foreign and sometimes inflexible set of procedures"* which *"did not adequately support multiple concurrent operation"*, *"predefined*

parts did not give the designer the freedom to work out new details that were unanticipated in the original system", and *"A building component based model is inadequate if it cannot support the open-ended needs arising from new, special details and from distinct or problematic site conditions."*[9]

GDS was a subsequent feature rich drafting system intended to compete with AutoCAD: however, AutoCAD supplied "80% of the functionality at 20% of the price."[10] GDS was aimed predominately at GIS and Facility Management markets, with a large installed base of loyal users. McDonnell Douglas bought out ARC in 1984 and took over the worldwide marketing of GDS, but following financial troubles in the aircraft division, GDS was sold to an arm of GM where it was disbanded soon after their takeover in 1991. Current GDS II software is unrelated.

HARNESS

The UK Department of Health and Social Security funded the development of HARNESS through Cambridge University School of Architecture Research. They set up the CLUBFS and subsequently ARC, with the aim of automating the design of hospitals by

---

9   "Building Product Models: computer environments supporting design and construction", Charles M Eastman, CRC Press, 1999

10 Old adage

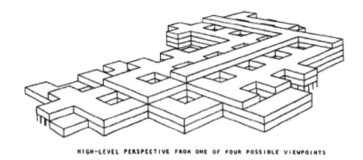

HIGH-LEVEL PERSPECTIVE FROM ONE OF FOUR POSSIBLE VIEWPOINTS

Figure 30. HARNESS Perspective from one of four possible perspective viewpoints from "Computer Applications in Architecture", John Gero, Applied Science Publishers 1976 © Elsevier

arranging standardized, pre-designed departments along a circulation spine. Complete design automation was achieved by standardizing the component parts. The system performed structural, environmental and cost evaluation, and was later used to manage the hospitals as well.

ARK-2

In the early 1970s, ARK-2[11] was developed by the firm Perry, Dean and Stewart in conjunction with two technical companies, Design Systems and Decision Graphics, based on a PDP-15 (1969) with Versatec electrostatic printer/plotters and Calcomp plotters. ARK-2 was a two-dimensional system using a refresh screen, limiting the complexity of the drawing that could be shown. *"Percy Thomas Partnership report that they can use up to 100 standard elements in a single drawing."*[12]

The system was designed to lay out space within buildings and had a number of programs dealing with the different aspects required, typically prefixed COMPRO:

COMPROGRAPH databases stored spaces from the model and were used by all programs in different phases of the design. They included space dimensions, ceiling heights and costs per square foot, or BoQ rates, and could generate both graphical and numerical reports (see Figure 26).

COMPROPLAN produced a bubble diagram of inter-related spaces that could be manipulated by the architect to produce various schemes. These could be exchanged for resized rectangles to give closer representation to the finished building.

COMPROSPACE took the rectangles of COMPROPLAN, giving the user the ability to produce drawings around the generated layouts. COMPROSPACE was a general purpose drafting program, using any stored drawing or plan, adding elemental and graphical data. *"Modification of the drawing can consist of moving, rotating or deleting any drawing element, adding a standard graphical element such as details, furniture and equipment from the computer maintained library."*

---

11 "Computer Applications in Architecture "- John Gero Applied Science Publishers 1977 page 312

12 "Computer Methods for Architects" R.A. Reynolds Butterworths 1980

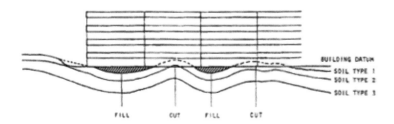

Figure 31. HARNESS Section through a building with automatically generated cut and fill from "Computer Applications in Architecture", John Gero, Applied Science Publishers 1976 © Elsevier

COMPRORELATE produced optimal arrangements of spaces and was able to generate square rooms that could be further manipulated to arrive at floor plans.

COMPROSPEC helped generate project specifications and COMPRONET performed critical path networking for project scheduling.

COMPROMAN, COMPROLINK and COMPROVIEW provided office management facilities, scoring facilities and producing 3D view perspectives of the building respectively.

EDCAAD OR SSHA

This project was developed by the Architectural School at the University of Edinburgh between 1969 and 1973, funded by the Special Scottish Housing Authority. There were two parts to the research, the development of a housing design unit with floor plans, and the development of site plans for housing estates.

Elements used to define the floor plan, such as windows and walls, were generically called 'blobs,' and were used to define the areas covered by each room. These were defined as part of the system. A wall "blob" carried material and other information associated with it. Room boundaries were automatically detected so floor areas could be calculated and surface areas of walls determined from heights of the walls. The system also had some capabilities to determine beam-sizes and heat-transfer, but the second part of the system was used for site planning. Unfortunately, the project leader, Aart Bijl, later concluded that the *"task-orientated knowledge implemented in the two systems was severely limiting, both in its generality and conception of architecture."*[13]

CEDAR

CEDAR 2 was developed by the Department of the Environment for the design of Post Offices in the early 1970s. The system enabled detailed design framing and external walling, from which cost estimation, daylight, thermal and acoustic data was derived. CEDAR 3 was a rule-dimension based system, with sets of preferred components and standard details. This system compared building and running costs for different building layouts. (The origin of CEDAR 1 is unknown).

13 "Building Product Models: Computer Environments Supporting Design and Construction", Charles M Eastman, CRC Press, 1999

GABLE

GABLE CAD was an advanced 2D and 3D design package with different modules, and was operated via a windows style interface and mouse running on UNIX. It was possible to create detailed 3D models and then generate 2D drawings or rendered visualizations from the data.

GABLE was developed at the University of Sheffield in the mid 1980s under the leadership of Professor Bryan Lawson, 2D Integrated Drafting System (IDS) generated 2D drawing files, allowing the user to define typical Autocad type drawings. The 3D system enabled the generation of 3D objects and captured different types of views of them. This was a separate database but the drawings it generated could be passed to the 2D system. There was also a separate database associated with the building, the Data Management System. 2D space models could be generated from this database and passed to the 3D part of system. GABLE was withdrawn in 1996.

CARBS

CARBS, Computer Aided Rationalized Building System by the University of Liverpool was developed in conjunction with Clwyd County Architects in 1972. Used for evaluation of plans and generating documentation, it seems to have been abandoned in the move to BDS-based technology.

SPACES

ABACUS, a unit at the University of Strathclyde, developed SPACES in 1972 to assist with the early stage design of schools. SPACES-1 performed spatial-analysis of the proposed school, based on subjects taken, numbers of pupils in each age group, the number of hours and the numbers of groups. SPACES-2 took the generated schedule of accommodation from SPACES-1 and created a bubble proportional to the size of the area in the schedule. The user manipulated these bubbles on the screen and converted them into rectangular spaces. SPACES-3 was used to evaluate the layout from SPACES-2. Data related to construction, activities and dimensions were added manually at this stage. Both CEDAR and HARNESS systems used similar approaches to the allocation of space, generating drawings automatically.

RUCAPS

RUCAPS was developed by Gollins, Melvin, Ward Partnership (GMW), a London firm of architects, where it was used for parts of

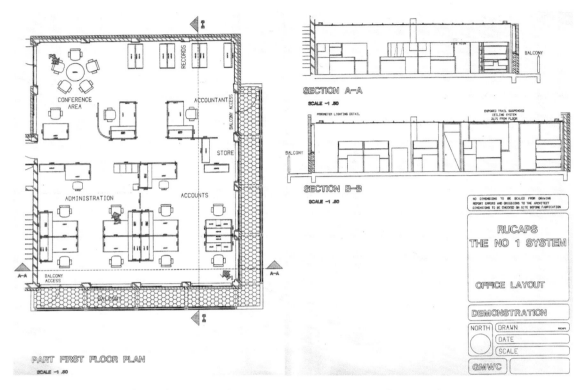

Figure 32. *RUCAPS Drawing Plan and Coordinated Elevation Demonstration 1983 © RIBA Archive, Victoria and Albert Museum*

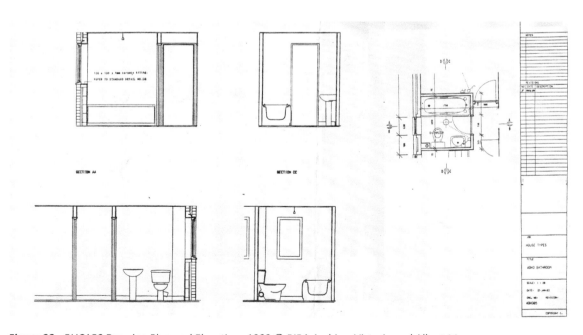

Figure 33. *RUCAPS Drawing Plan and Elevations 1983 © RIBA Archive, Victoria and Albert Museum*

the then new Riyadh University, and subsequently for various other projects, including the design of a number of large hospitals. Reich Hall Architects, Scotland, was one of the more prolific customers building several hospitals, mainly in Scotland.

In 1984, RUCAPS was described by Reynolds[14] as a *"2-1/2 dimensional interactive system, closer to the philosophy of 2D interactive systems such as ARK-2 or DAISY-1 and concentrating on the rapid production of 2D drawings (plans, elevations and sections)"*. Reynolds goes on to say *"Supplementary results are not considered important to RUCAPS but simple scheduling is available"*.

RUCAPS carried the 'weather vane' or 'box geometry' concept from UK-based Applied Research's Building Design Systems to generate coordinated 2D drawings. As stated earlier, this philosophy is now found in most modern BIM systems and was used in SONATA, REfLEX, and more recently Revit. Early ArchiCAD carried a plan and perspective view.

The Author worked for GMWP (later GMWC) on RUCAPS and added an external 3D database to RUCAPS so that 3D images could be rendered and provided a rendering algorithm and hidden-line algorithm. A separate 3D file of information was generated from the scheduling program and that was read into a separate program which added 3D data from manually generated files, to make a static "flat" file of vertices, lines and faces. This was then passed onto the hidden surface program, separating the 3D view from the 2D model. It was not possible to work in the perspective views and there were no links from the 3D back to the 2½D model. Algorithms generated images and displayed them, but changes in the RUCAPS model were not reflected in the 3D data unless the entire process was repeated. The brilliant Jeff Findlay worked with the Author to produce the movie "Esquisse" with this "bolt-on database," taking over 1000 hours of compute time, 6pm till 9am over a 9 month period, unlike lightning!

The Author also implemented clash detection as part of the hidden line removal calculation. This flagged any "inter-penetrations" that occurred between any two 3D shapes in the 3D file.

RUCAPS had no 3D Boolean operations. Similarly it was neither possible to work in 3D nor manipulate 3D objects graphically. In particular, there was no "joining" of different wall and windows/

doors types, all found in modern BIM systems. Hence, in RUCAPS, if one placed a wall and a window in plan, one had to manually break the wall into two parts in order to insert the window. The window had to contain separate wall parts above and below to repair the break in elevation. The system had no idea that windows and walls were joined. Schedules etc., did not reflect the fact that the window was in the wall, reducing areas required. Cavity closing details were inserted by hand later.

RUCAPS did not have parametrics per se, however a video of the time talks about parametric elements but in fact no elements were shown. The Author has never seen any evidence of parametrics of any form.

So according to the BIM Handbook[15], systems that do not support parametric behaviour are not BIM. The separate 3D file (implemented by the Author) is non integral and is also not BIM

Aishe and Bredella in their conversation document "The evolution of architectural computing: from Building Modelling to Design Computation"[16] refer to RUCAPS as one of the earliest "building modelling" systems (as opposed to BIM). As we have seen, there were a number of earlier building modelling systems (as opposed to BIM). Aishe also refers to RUCAPS as being proto-BIM. The previous paragraphs indicate how RUCAPS lacked many of the features essential to BIM. This includes the single model (if one counts the separate 3D model that the Author added), parametrics components, multiple person access and so on. RUCAPS was a step on from BDS, but as we have seen, there are many other components to making a BIM system.

All RUCAPS development ceased when GMWC purchased SONATA from the Author in 1988.

GLIDE II

GLIDE was developed by the CAD Graphics Laboratory at the Carnegie-Melon University under the leadership of Charles Eastman. Quoting Eastman, *"the aim of the project was to develop a language for 1) the development of solid modelling for use in building, and 2) the integration of solid modelling with database and other capabilities needed to generate an environment for*

---

14 "Computer methods for Architects", Reynolds R.A., Butterworths, 1984

15 "BIM handbook", Eastman et al., Wiley, 2018, page 20

16 "The evolution of architectural computing: from Building Modelling to Design Computation", Robert Aishe and Nathalie Bredella, Architectural Research Quarterly Vol 21, Issue 1, 2017

advanced CAD system development."[17]. This language was Interpretive, an extension of Pascal, and it formed the basis of BDS (not to be confused with the identical acronym Building Design System BDS from Cambridge discussed above).

Building Description System, (BDS-USA) was solid modelling-based, released in 1974, developed to form the Graphics Language for Interface Design (GLIDE), a general-purpose language for designing Graphical User Interfaces (GUIs). It interpreted an extension of the Pascal language, popular at the time, and enabled Boolean geometric operations on building descriptions.

CAEADS

The University of Michigan Architectural and Planning Research Laboratory investigated Computer Aided Architectural Design (CAAD) from the early 70s and developed the Computer Aided Engineering and Architectural Design System (CAEADS). This was led by Harold Borkin and supported by the Army Corp of Engineers Construction Research Laboratory and was used to design military facilities. It is an integrated set of automated tools to support data development in conducting studies in the early concept design process. CAEADS programs interface with other standalone programs, such as energy analysis, structural analysis and drafting systems.[18]

CAEADS designers took the approach that building design is not about drawing, but rather about modelling. "*Designs must be based on some type of meaningful analysis. Therefore a system for computer-aided design must have broad analysis capabilities. The system should not be based on computer graphics, but on computer modelling. If the building is properly modeled, the production of drawings is only one of the many uses of the model.*"[19] Following this notable concept, CAEADS attempted to integrate a range of individual engineering programs for different types of analysis, most of which were stand-alone, punched card based, so it was a difficult approach. A number of analysis programs were implemented using a common data file, and each system added its own file structure to save relevant data, and ran only on University of Michigan's mainframe computer. The School of Architecture of New Jersey

Institute of Technology also used CAEADS to simulate human behaviour in buildings.[20]

We have seen various systems that assisted in the design of buildings. These systems were expensive, mainly because the computers were immensely costly. This limited the market of these systems. Around the middle of the 1980s machine process fell and the market opened up to a wider ranger of vendors. These are categorized below.[21]

CALCOMP

Calcomp produced ADP and AVP, Architectural Design Package and Architectural Visualization Package respectively. ADP produced 2D drawings and AVP perspective and isometric drawings from the plan or elevation views. Changing a plan, or elevation, meant that the 3D had to be regenerated, and even Plan and Elevation drawings may also not be consistent between themselves; a further step was required by the user. It was very successful in its time.

ARRIS

Sigma Design Inc of Denver developed, marketed and supported ARRIS, a 2D system with bolt-on 3D parts. Many of the modules were written by 3rd party developers using a common interface, but the data transfer between the modules was not always straightforward. FM was their major market. Sections and elevations were potentially unrelated to the plans.

ARCHITRON

Architron, from the French company Gimeor, was an Apple Macintosh-based 3D CAD system. 2D drawings were generated from the 3D model, regenerated each time, and was one of the best architectural CAD systems on the Mac but suffered from the restrictions of the Mac[22]. It was a 3D system with added 2D. There was no data associated with elements, giving limited parametric functionality, again successful in its time.

17 "Building Product Models: Computer Environments Supporting Design and Construction", Eastman, C., CRC press 1999

18 "Computer-Aided Engineering and Architectural Design System §Test Results", Spoonamore and Golish, US Army Construction Engineering Laboratory

19 "CADIA at Michigan", Theodore Hall, CADAIA, Dec 1983, Vol III. No 2

20 "Simulation modeling of human behavior in buildings", Filiz Oziel, School of Architecture, New Jersey Institute of Technology, University Heights, Newark NJ

21 "Alias SONATA Sales Guide", Alias Research

22 The 1984 Macintosh was the first commercial system to use pull-down menus and icons. It had a memory of 128KB (increasing to 1MB) and a screen resolution of monochrome 512x342 pixels.

ARCHICAD

Founded in Hungary in 1984, GRAPHISOFT produced a 3D CAD system ArchiCAD for Architects and a comprehensive parametric 2D drafting system which they sold in parallel until the early 1990s. Initially released on an Apple Mac in 1984, a Windows version was released with V4.12 in 1993. ArchiCAD became available with added parametric 2D with V4.5 circa 1994, and with a "file-based teamwork solution" (multi-user) in 1997, making it integrated 2D and 3D from that time

It was similar to other 3D systems that generated 2D views from the particular projection of the 3D building model. The 3D model was used to generate the 2D projects that were then annotated.

In 1988 they had also released a 2D drafting system TopCAD, sold alongside ArchiCAD until the early 1990s when ArchiCAD functionality included that which TopCAD covered. According to the Hisrich document, *"TopCAD is a high-end 2D CAD solution on the Macintosh for users who need precision, advanced editing, associating dimensioning, 2D parametric and other features."*[23]

Graphisoft's ArchiCAD is one of the most successful BIM systems available today and it has also been suggested that it was the first BIM system. In an email to the Author on 21 August 2016, Gábor Bojár, the founder and current Chairman of Graphisoft, has put this into context which he has agreed may be published as follows: (all italicised text in quotes are direct quotes from his email).

ArchiCAD pre 1995 was a "predecessor of BIM" for various reasons:
"Only on the floor plan view were you able to enter data or modify anything on the model" and *"After any changes you had to regenerate the entire 3D model in order to see the changes on the other views (for more complex models it was rather time consuming)"*[24]

While the 3D GDL view could be changed, the graphics in the elevation and sections view did not change unless the updating process was complete.

Graphisoft's founder and Chairman, Gábor Bojár, interviewed in 1990 by Professors Robert Hisrich and Janos Vecsenyi stated

"ArchiCAD had a separate 3D and floor plan symbol for objects from the very beginning. It was literally a drawing and could only be stretched proportionally by its overall sizes (A,B parameters), while 3D was parametric."

Quoting Gábor Bojár once more from his 2016 emails:
"The drawings were not detailed enough to be working drawings (for example there were no cavity walls)" and *"The early versions could not handle large and complex models, primarily due to computer power limitations."*

The first "multi-user" release of ArchiCAD appeared in 1997 with V5.1 and symbolic views and dimensions were added after drawings had been regenerated.

See also Appendix 6 for a letter from Gábor Bojár to the Author.

IBM

IBM produced three CAD packages in the 1980's: CADAIM was used for manufacturing, CATIA for 3D Modelling, mainly mechanical drafting, and FASTDRAFT, was a 2D drafting system.

AES

Architectural Engineering Series (AES) was bought out by Skidmore Owings & Merril (SOM), the largest architectural engineering practice in the U.S. at the time, and marketed by IBM on their behalf. Between IBM and SOM the package had huge credibility, but in spite of this, AES did not do well with the exception of Denmark, so with the 1987 recession and lack of sales, SOM downsized the development team. The package offered in the U.S. included several engineering design and analysis modules. Still, the system sold in the rest of the world only included separate 2D and 3D modules as their engineering standards were considered to be different. IBM contributed to the product's downfall by selling it through dealers who had little or no knowledge of the applications.

INTERGRAPH/BENTLEY

INTERGRAPH, founded in 1971 by Jim Meadlock, produced hardware for graphics workstations, IGDS. They launched a 3D plant design system in 1985, which formed the basis of Microstation's file format, the PC-based CAD product owned by Bentley Systems. IGDS allowed 2D and 3D file structure but was not structured so that changes to one file would change a separate drawing file. PseudoStation, the first product produced by Bentley in 1984, had initially allowed users to view IGDS files without

---

23 "The Entry of a Hungarian Software Venture into the US Market", Professor Robert Hisrich and Professor Janos Vecsenyi, The European Foundation for Entrepreneurial Research, 1991

24 Email from Gábor Bojár to the Author 21 August 2016

using INTERGRAPH software to edit the IGDS files. INTERGRAPH subsequently purchased 50% of Bentley.

COMPUTERVISION

Computervision (CV) was founded in 1969 and in 1977, they introduced their own computer and manufactured a new workstation. Their product CADDS-4 was introduced in 1980/81, and during the period 1973-83, CV maintained a European Sales record of 60% growth per annum.

During this time, CV was the world largest supplier of CAD. It was a closed system until 1983, making it very difficult to write new software or offer 3rd party software for the workstation. CADDS-4 solved this, and the system went from strength to strength, with CV acquiring a number of companies including CIS, GRADQ, UNIRAS, ALDPIPE and SUN. The result was that instead of a closed turnkey solution, a much wider range of hardware and products was available. The resulting product CDS300 was a group of separate packages for drafting, space planning and preparation of technical publications, where optional features included a "repeat generator", spreadsheets and word processing. CV was acquired by Prime Computer Inc. in 1988. Prime changed their name to CV and was subsequently acquired by PTC in 1998.

WHERE DID THE BIM PHILOSOPHY ORIGINATE?

All of the systems characterized above were used to design buildings. Various universities had spent a decade or more looking at the problem of building design and systems varied, each having its own unique approach from different people working on similar complex problems.

A common feature of many of the pre-1984 systems was that of automating the design process. The automated outputs, typically generating room layouts that could be extruded into simple 3D shapes, allowing 2D drafting information to be added. Many specialist buildings were designed and built using these design tools. These systems involved building layers of information and hierarchy enabling a specialist design to be created. This type of automatic design system was abandoned with BDS (ARC) being perhaps the last. RUCAPS was an exception to this group in that it concentrated on coordinated drawing production.

Subsequent systems took to modelling in either 2D or 3D with a view to producing those drawings. They sometimes allowed the other types of information, 3D or 2D respectively, to be generated.

For instance generating 2D drawings (but not symbolic) from the 3D model.

But none of the early systems managed to encapsulate Chuck Eastman's insightful ideas of that time. He wrote:

"It would combine the positive aspects of both drawings and models and eliminate their common weaknesses. It would incorporate three-dimensional information … and would require any changes to be made only once for their full effect to be revealed. It would accept changes easily and provide automatic checking for spatial conflicts.

A building can be conceived through a collection of three-dimensional elements arranged in space… Design would consist of interactively defining elements according to their shape and other properties, and arranging them much as one would arrange a balsa-wood model. It should be possible, then, to derive sections, plans, isometrics and perspectives from the same model. Any change of arrangement would have to be made only once for all future drawings to be updated. All drawings derived from the same arrangement of elements would be automatically consistent."

SONATA was the exception with integrated 2D plan, section, elevation and 3D treating all types of views equally. It also integrated information into a single parameterized database, together with full building information and pointers and structures for the system to understand the structure of the building.

SONATA

In the early 1980s, the Author wrote the SONATA system alone in an attic. Most of the images in this book are from the Author's two systems SONATA and later REfLEX. A brief personal version of the creation is given in the Introduction and Appendix 1 describes some of the technical content. Appendix 4 was produced to try and raise capital to start the process and Appendix 5 was produced after completion. The Appendices are copies of the original documents.

Quoting from these documents:

"Developing a computer aided design system for architects, engineers and other professionals involved in planning, design and constructing buildings."
"It uses a single model to fully co-ordinate graphical and non-graphical information of all objects within the building."

"...'intelligent' components are able to design and modify themselves depending upon the surrounding conditions in the model."

"... other design facilities include network design facilities, structural steel design, thermal analysis and so on."

SONATA meets all of the criteria described by Eastman[25] for defining BIM.

SONATA has a single building model into which, and from which, information can be inserted and retrieved. Components are assembled onto layers set at different datums in the building. The user can work in plan, elevation, section, and 3D, singly or all at once. Changes in any view are reflected in all views. Drawings are extracted from the model at the moment they are needed.

Automated coordination of drawings, 3D and schedules, parametric capabilities for components to self-design and communicate information, in addition to information passing to allow design information to be passed, were all self-evident as issues that needed to be solved and automated. Joining walls so that the schematic view in plan had cavity closure, creating holes in the wall in 3D, drawing the appropriate sections were all issues that could be addressed by passing data between the different items. Wall joins in plan elevation and 3D and reflected in the schedules were also addressed.

---

25 "BIM Handbook" Eastman, Teicholz, Sacks and Liston. Wiley 2018

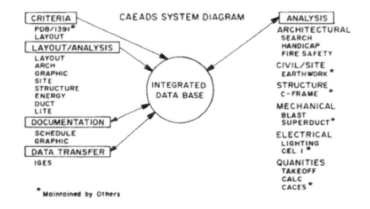

South West Perspective

Figure 34. *CAEADS Drawing Education, Training and Employment Facility 1984 from "Computer Applications in Architecture", John Gero, Applied Science Publishers 1976 © Elsevier*

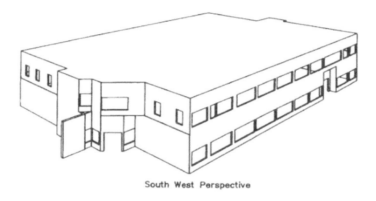

Figure 35. *CAEADS Implemented analysis programs from "Computer Applications in Architecture", John Gero, Applied Science Publishers 1976 © Elsevier*

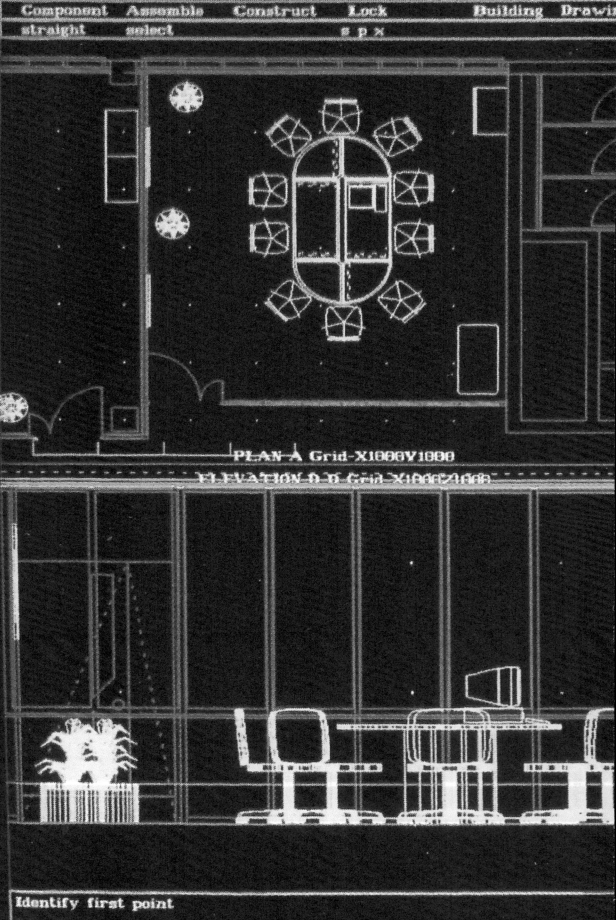

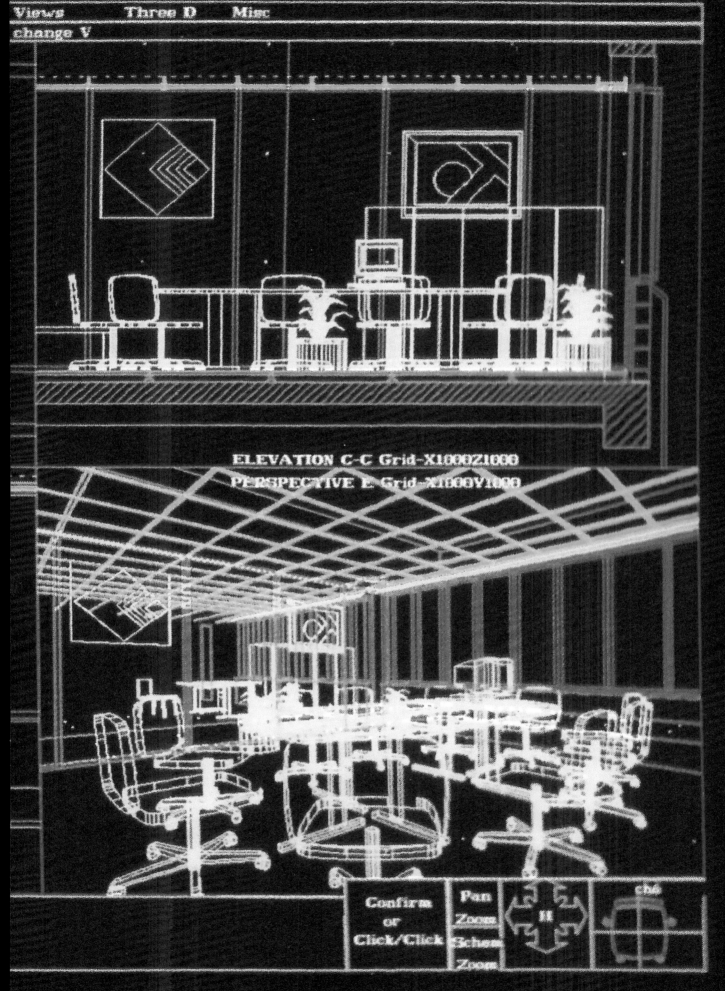

Dragging an object in one Window would result in the same object being dragged in the other three. Also note hatched section objects and mixture of symbolic and 3D. Hole cut by door in wall is generated each view. 3D points can be selected between the different windows.

Defining drawings and sheets, 3D views and families defined by categories all developed in order to ensure drawings were correctly regenerated.

In constructing the new type of model, the primary aim was to be able to answer all of the questions and demands placed that might be asked about the building to be built. This was achieved by allowing

- Fully user defined parametrics in all plan elevation section and 3D component views, which had not previously been achieved by any other system until ArchiCAD achieved this in 1995
- Full Boolean shape operations in 3D to allow cavities in walls for windows and doors and in slabs as well as 3D shape generation
- Automatic wall cleanup/joins and window/door insertions with full calculation capabilities
- Bi-directional working in plan, elevation and 3D windows simultaneously
- Time based construction
- Drawings and sheets automatically regenerated
- Automatic animation sequences
- Automatic solution creation and solving of physical networks
- Proprietary rendering, colour change, spot lights, shadows and other effects
- In addition to all the normal dimensioning, fonts hatching and other normal drafting activities

REFLEX AND PROREFLEX

The Author then went on to write REfLEX, with the help of several team members who had become involved with SONATA.

Designed and completed between 1992 and 1993, REfLEX was effectively the son of SONATA. REfLEX was constructed to include all of the SONATA functionality, but with new approaches on several fronts.

REfLEX had several main components, the Modeller that allowed fully coordinated project modelling, the Renderer that allowed photo realistic images, *"including the effects of transparencies, textures, displacement maps and multiple light sources with shadow casting."*

A third party renderer was also included called "REfLEX Reality" which included ray-tracing with reflections, refractions, smooth shading, and soft shadows.

In addition to the basic engines and the company itself, REfLEX built libraries of components specific to the different construction industry disciplines. Third parties provided some of this software. Application Libraries are described in contemporaneous documents as:

REfLEX Architecture, a comprehensive architectural element library. All elements have 2D and 3D views, most have sectional and elevation views, and several elements like doors and elevators are operable or have moving parts. It was able to switch lights on and off, open and close doors and window blinds.

HVAC Ductwork library contained all the elements for duct layout and design, all parametric (impossible to do without parametrics). Elements use link point functionality in REfLEX, allowing section, size and finish to be automatically passed from one duct to the next. This library was developed by Engineering Technology Ltd. The Structural Steel library had a full set of elements for the design of steel frames including a wide range of cross-sections.

REfLEX was sold to Parametric Technology Corporation (PTC) in 1996. It was marketed as ProReflex. This was a direct derivation of REfLEX and SONATA, using the same source code on UNIX and Windows NT in 1997. Development was stopped and several employees, the founders of Revit, were given a "non-exclusive development license" by PTC for the ProReflex software. This included the sources, tuition and full development rights. The Author believes the founders of Revit used detailed REfLEX intellectual property and key knowledge to develop Revit. Almost all of the functionality of REfLEX can be found in Revit. Also many hundreds of thousands of man-hours went into the development of SONATA, REfLEX, ProReflex and the earlier generations of systems and it is unlikely that Revit developers reinvented all of this from scratch without reference as claimed.

BIM systems have all used this integrated, parametric, information rich, structure aware database. This includes SONATA, REfLEX, ProReflex, ArchiCAD, Revit, and VectorWorks.

Current BIM Systems 4

In the first chapter, we have examined the issues surrounding constructing structures of complexity, and we have looked at generic BIM as a solution to these problems. Here we explore some of the different commercial systems that are available today.

Perhaps stating the obvious, the particular system and its implementation are very relevant as to overall performance. As in many things, there are at least 50 shades of BIM. Some systems achieve one part, and others address another part, and so the choice is not always clear cut. One of the problems is that some vendors say this is BIM when, in fact, what they have is a 3D modeller.

To achieve maximum BIM effect, the software and the hardware choices are crucial. There is a mantra that BIM is not the software and that it is a process. It is both. If you don't believe it, try BIM with weak software. The software is crucial. It enables and enforces sharing, and it performs the essential updates across the model. It is hard or easy to use. Let us look at the different systems and see what is possible.

Software development is an expensive operation, and so when charged for, is seen as a barrier to BIM. Perhaps apocryphally, 1 in 20 of ALL software projects make it to market. So the writers of these systems need to charge. Eventually, the cost of the software will fade into insignificance when compared to the cost of management change.

Some of the BIM software vendors have chosen to make the software a recurring cost. The motivation is primarily share price, but the answer given is that it pays for improvements and support. But try and say you don't want improvements or support and you will not be able to buy at all, not for a single up-front price. Gone are those days.

The well known Moore's law states that computers double in speed every 1 or 2 years. Unfortunately, the trend in software today is to use every cycle of computing power available, so effectively Moore's law benefits are lost. Even in the 1980s buildings of considerable complexity were designed and built with BIM. Every multiplication and division were counted to ensure that the programs were as efficient as possible. This was done as the machines ran many thousands of times slower; if you didn't, the process would take years. So the moral of this is to buy the most powerful machine you can. The software developers and their tools will use all the power you can provide.

One would expect the size of models that can be tackled to grow as the power and size of machines has grown. Can you do a project that is many thousands of times larger than Figure 180? This figure is a full BIM model, with connectivity between the various ducts and pipes and obscuration between layers. One might expect to as that complex model was done on a machine thousands of times slower and smaller. Space requirements have also grown significantly. Revit tends to expand the BIM database almost catastrophically, especially when using families. This will very quickly lead to a file size that is huge and potentially slow.

To give BIM the best chance to work, we must also choose a system that causes the least resistance to implementation, allowing the different stakeholders to work together. This must be at a sensible cost. What that cost is will depend upon your organisation.

Most of today's major software providers allow users to try the software for free, so try these systems, read up about the technology: there is no shortage of information available online; get a feel for what is needed.

Below is a review of some of the BIM systems. It has been written without prejudice or bias as the Author has no affiliation with any of the vendors.

It is worth remembering that BIM is collaborative and the importance of considering what everyone else is going to be using. Using files such as Industry Foundation Classes (IFC) is not a good solution as immediately you have a copy of the model that will rapidly go out of date. Multiple copies take us back to the good old days where you were never quite sure which IFC files were up to date. So choose a system everyone else is running if at all possible.

This suggests that you use a system that the other stakeholders in the project are using. This will probably mean Autodesk's Revit. Sharing models between the different system software suppliers is fraught.

CURRENT BIM SYSTEMS

There is a wide range of "BIM" software systems available today each with their view on BIM. The actual number is upwards of 70 and so the offering below is a small selection of the better known. Not only is the market large, it is also changing rapidly making reviews very difficult. Among the market leaders are:

- Autodesk Revit (Architecture/Structures/MEP)
- Graphisoft ArchiCAD
- Nemetschek Vectorworks
- Bentley OpenBuildings
- Beck Technology
- Nemetschek Allplan
- Bricsys BricsCAD BIM

There are a number of other BIM systems available including Trimble owned Gehry Technologies Digital Project, Tekla Structures and Vico Software Constructor. These have relatively small market shares and are not considered here.

Some of the systems are ideal for sketching simple house models in 3D. Be wary of systems that cannot deal with large models, cannot deal with services, and of systems that do not deal in an intelligent way with data associated with the Objects. Some of these have been excluded from this list. Others have been excluded because they are very specific to a single purpose.

REVIT

Revit is the leader in the BIM software market in most countries with Austria, Germany, Switzerland and Japan being exceptions. Charles River Software, founded in 1997, developed Revit after obtaining a non-exclusive development license for REfLEX from Parametric Technology in 1998. The company changed its name to Revit Technology Corporation in 2000 and was acquired by Autodesk in 2002. In addition to obtaining a development license from Parametric Technology (PTC) of the ProREfLEX system, the arrangement included the source code and access to the REfLEX/PTC staff.

As with its forebears, Revit is a parametric modeller with an instance model of the building. Instances of components, representing the structure and their inter-relationships are created and stored with associated information in a central database for each project. All information is automatically updated with changes guaranteeing an up-to-date model. It can be viewed and changed by any team member, and these changes are reflected throughout all accessing programs.

Quoting from the front CD cover for version 1 of Revit published in 2000, "*Revit, revolutionising building design*" and "*The first parametric building modeler.*" Let the reader determine the truth of this.

Autodesk provides a wide range of file formats including DWG, DWF, DXF, DGN, SAT and SKP files and for video and rendered images JPG, TIF, PNG, AVI, TGA and BMP. Revit also supports several standards for structural analysis, energy simulation and load.

Advantages
- Large market presence so easy to share model & components
- Large development team
- Strong development platform (lots of add-ins)
- Widely supported with courses, books, tutorials,
- Visual scripting through Dynamo.
- NET scripting
- Manage drawing revision sets

Disadvantages
- Does not run on OSX
- Only partially multi-threaded
- Needs fastest CPU
- Slow on large models
- Does not scale well to very large models
- Revit Data files is not backward compatible and is forward compatible with mixed results if too many versions jumped

- projects grow disproportionately when using multiple families
- Models get too big to load

Families are a contentious point in Revit. At first sight, the fact that relationships are created automatically between different groups of elements is useful; however, it turns out that the built-in relationships and constraints can sometimes be a hindrance. Walls, floors, columns and ceilings have relationships to each other based on their relative positions. This can cause the model to be inflexible during the evolution of the design. They create dead-ends that cannot be anticipated but can only be resolved by deleting elements and rebuilding them in the "right" order.

The Revit interface is somewhat more complex than either Vectorworks or ArchiCAD. It tends towards menus within menus.

Revit has the advantage when working on buildings with many floors. In ArchiCAD, you need to have every floor constructed; in Revit, SONATA and REfLEX, the principle floors can be referenced without reproducing the detail. This simplifies larger projects.

One of the downsides of using a family or a graphic editor is there is no ability to add complex external functionality or plugins. In SONATA and REfLEX, the user could program (albeit with programming language) plugins that enabled many different types of functionality. All system variables from the programs themselves were available, and all of the parametric variables, as well as variables in adjoining parametrics.

What this means is that the user could add functionality into the objects or model. REfLEX and SONATA could do things as diverse as reading external system files and sensors, solve networks of objects, control external processes, generate live reports as the model monitored the real world, and produce reports and drawings that changed in real-time.

GRAPHISOFT ARCHICAD

Graphisoft was founded in 1984 by Gábor Bojár in Hungary developing a product called RadarCH, "3D designing software on a personal computer" for architects. The Graphisoft 2D drafting program TopCAD was sold in parallel with RadarCH from that time. This program had comprehensive 2D parametrics and was for users who wanted a high-end 2D drafting program with "precision, advanced editing, associative dimension, 2D parametric and other features normally found in workstation or mainframes". ArchiCAD was derived from RadarCH and by the mid-1990s V4 .5 was parametric across all views allowing most of the functionality of BIM.

ArchiCAD has had regular releases since the mid-nineties and holds a significant market share. Influenced by SONATA, it is built around the concept of a "container of views" or the concept of multiple views attached to an object. ArchiCAD offers an extensive range of objects and a very powerful language to describe the different views of the components.

A fundamental difference between Revit and ArchiCAD is that Revit uses a graphical construction technique to ensure that the different views of the component, the plan, the elevation section and 3D have a better chance of matching. ArchiCAD allows the user a bit more freedom and a lot more power, to write the views in a scripting language. This has been quoted as a disadvantage but in fact, ensuring the views of a component match is trivial. The power the language brings is enormous, allowing iterations, internal design calculations, and so on.

Initially aimed at architecture ArchiCAD has now evolved into a full BIM system with its MEP modeller. Graphisoft also produces its own MEP Object library and Toolbox. The library contains an extensive list of MEP specific objects or components. Structural design is done via external systems such as Tekla using IFC to communicate the specific elements. ArchiCAD can export the standard image formats (JPG etc.) and import and export DWG and DXF files (2D drawing files).

Advantages
- Fully multi-threaded
- Runs on OSX and Windows
- Supports McNeel Rhino and Grasshopper for complex computational generation of geometry
- Has a friendlier user interface (not dialogue after dialogue)
- IFC well supported
- Custom Property Sets added easily
- Very flexible and efficient parametric objects (through GDL scripting)
- Smaller files
- Handles more complex geometries

Disadvantages
- Custom objects require GDL scripting
- Many old workarounds were never fully resolved
- Some extensions were never properly updated (most notably: stair maker)
- Less-used API (C++ with limited documentation and examples)
- Mainly orientated towards architects

VECTORWORKS

Vectorworks has a strong presence in Europe but is sold around the world in over 75 countries. Diehl Graphsoft was founded in 1985 to undertake the development of a 3D tool, and Minicad was written. This was a 3D tool that produced 3D drawings that could be edited to produce construction drawings. In 1999, Minicad became Vectorworks having gained a range of architectural objects and parametric constraints. In 2000 Diehl Graphsoft Inc. was acquired by the Nemetschek Group of Munich, Germany and the name was changed to Nemetschek North America, Inc.

Vectorworks evolved into BIM in the modern sense of the word in the early 2000s and by 2010 had full parametric 2D/3D BIM capabilities. At this time it had the powerful capability of modelling in NURBS utilising the Parasolid modelling kernel, a capability that allows the user to generate complex, smooth surfaces such as you might find on a car or large curved building.

Vectorworks has an extension Extensio Pro for MEP plugins. This enables the creation of 3D duct, conduit and plumbing networks. Vectorworks allows communication with Scia Engineer for structural design, though this is an external package and is not integrated within the model. It also uses the Parasolid kernel; this is a powerful tool for lofting curved surfaces, Boolean operations and blending and filtering.

Vectorworks became active in the OpenBIM interoperability program and hence has IFC import and export capabilities. Similarly extensive (greater than everyone else?) import and export of pretty much every file format available, particularly DXF, DWG, EPSF, WMF, PICT, PDF, SHP, 3DS, IGES, SAT, SKP, X_T, JPG, GIF, TIFF, PICT, PNGT, and PNG.

Advantages
- Simple interface
- Intuitive
- Concept through design
- Revit Export
- Walk-through animations
- GIS integration
- Live data visualizaton

Disadvantages
- not Revit
- Basic landscape design

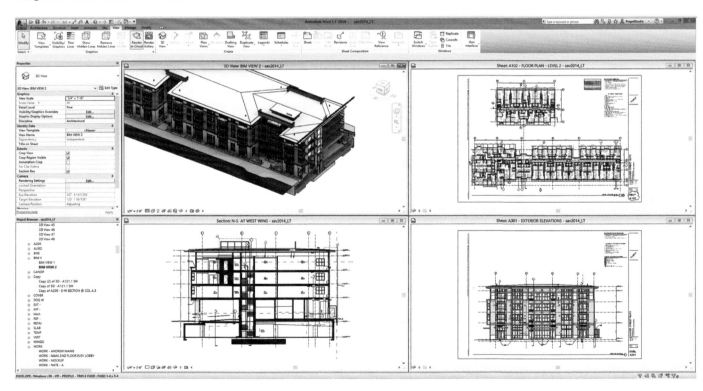

Figure 37. *Revit Screenshot Architecture 2014 © Autodesk 2019*

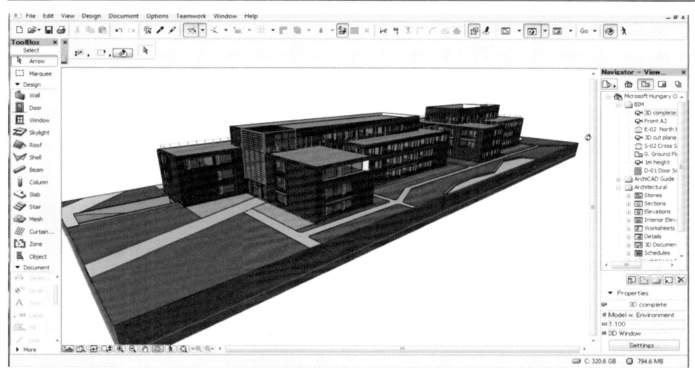

Figure 38. *ArchiCAD BIM model © Graphisoft 2019*

It provides high-level architecture, FM and town planning tools at an affordable cost, worthy of close examination.

BENTLEY OPENBUILDINGS

Bentley was a latecomer to the BIM market, introducing its first BIM-based software, Bentley Architecture, in 2004. Bentley's first products were MicroStation, a 2D drafting package that edited IGDS files, and TriForma (written by Bricsys and licensed), a 3D modelling package for Architecture both from the 1980s. They were combined to form Bentley Architecture with a rich set of editing and surface functions. Bentley Architecture V8i has structural, mechanical, electrical, FM and civil components, making it a fairly broad system. Similar to Revit, Bentley Architecture has a parametric capability.

Bentley's approach is different and in a way, is not pure BIM. The model is an Integrated Project Model but is in effect a series of programs working on a range of files, which without the correct programs maintaining it, is not a single model per se. This works well providing those programs are run.

Advantages
- Customizable User interface

- Multiple File formats
- Allows complex surfaces

Disadvantages
- Steep learning curve
- Needs full program set to guarantee BIM

BECK TECHNOLOGY DESTINI

Beck Technology Destini (design, estimating, integration and initiative) Profiler is a parametric modelling system used for the conceptual design of buildings. It provides accurate construction costs and estimates, energy costs, optimising different aspects of the building. The software is used to access the feasibility of the project early on in the development cycle and can be used as the basis of IPD.

The Beck technical lead comes from Stewart Carroll who had worked for REfLEX and later PTC as a software engineer in the mid-nineties. Beck purchased the intellectual property of ProReflex.[1]

---

1   "Integrating Project Delivery" Fisher, Ashcraft, Reed, Khanzode, Wiley

Figure 39. Competition entry for the Mali Museum in Lima, Rhino, using VisualARQ and Lumion 6.0. ©, Images and credits: Leandra C.Boldrini and Pedro O.P.Mariano Brazil

Figure 40. Competition entry for the Mali Museum, Lima, ©,Images and credits: Leandra C.Boldrini and Pedro O.P.Mariano Brazil

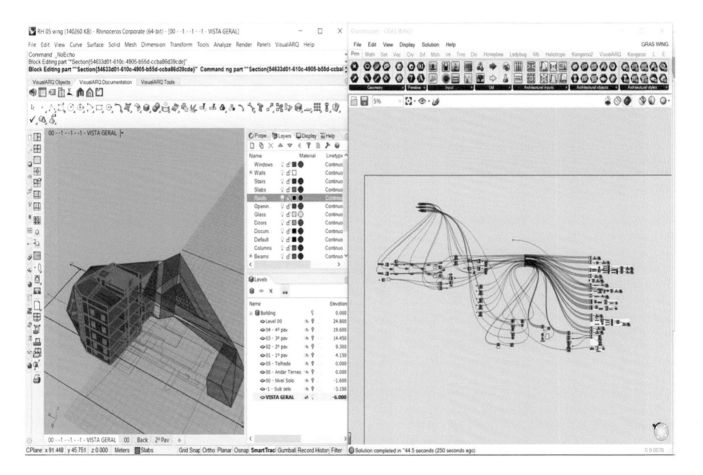

Figure 41. Using Rhino (left) and Grasshopper (right) to produce the lofted surfaces, © and credits: Leandra C.Boldrini and Pedro O.P.Mariano Brazil

There are three packages that are currently available: Profiler, Estimator and Optioneer.

- In Profiler, a simple 3D model is created giving feedback on costs, energy, cut and fill and schedules. It integrates Sage Timberline applications.
- Estimator provides information around the complexities of construction estimating, allowing different ways to organise direct and indirect costs. Estimator outputs to Excel, following the project as it is developed.
- Optioneer optimises different parameters across the building, early in the design process. Orientations, configuration and lowest capital cost are all available from this package.

Destini is limited to DWG, DXF, IGES and STL. Models can be output to IFC and Quest is directly integrated to give quick energy calculations. Profiler is not a general-purpose BIM tool.

BRICSYS BRICSCAD BIM

Recently acquired by London-based Hexagon, Bricsys is an old hand at developing systems. They developed Triforma which Bentley sold as its BIM application in the 1990s and later a tool called Architectural. BricsCAD is cheap, very cheap compared to the Revit's of this world. BricsCAD BIM, less so but still the perpetual license as opposed to annual fees forever. They have been adding BIM functionality, including collaboration together with powerful lofting[2] tools and some parametrics.

NEMETSCHEK ALLPLAN

Nemetschek Allplan offers a range of tools for architects and engineers merged into the BIMPlus tool. It comes with 3D modelling capabilities enhanced by Parasolid kernel. It allows data to be shared across all stakeholders and provides more flexibility than most. The normal additions of visualization (the excellent Maxon CinRender) and data exchange are included. Also worth considering.

RHINO, GRASSHOPPER & VISUALARQ

Rhino is not a BIM system, but rather a useful addition to a BIM system. It is a software tool developed by Robert McNeel & Associates. It uses NURBS (a mathematical surface) to generate general purpose free form surfaces It now includes Grasshopper (from V6 of Rhino), a programming environment that assists in this process.

---

2   Lofted Surfaces are surfaces generated by multiple 3D curves or existing edges creating a smooth tangency between the selected curves

Rhino includes a set of tools to create and manipulate surfaces, rather than polygon meshes, in a way that is amazing and simple, given the complexity of the mathematics that is going on behind the scenes. These tools include transform tools, points and curves options, mesh tools to export the meshes and tools to convert NURBS to meshes for rendering.

When used in conjunction with VISUALARQ these tools become very powerful in the design of free form buildings. VISUALARQ is an architectural plugin for Rhino, and uses IFC to transfer information between BIM systems and Rhino and Grasshopper. See Figures 39 through 41 opposite. This project was devised by the Brasilian architects Leandra Boldrini and Pedro Oscar, for the design contest of the new wing of the Mali Museum in Lima, Perú using VisualARQ, Grasshopper and Rhino. The results were rendered in Lumion 6.0.

Grasshopper is a node-based, visual programming environment and is used within Rhino. It enables you to create parametric structures within Rhino, being able to view and manipulate them. You can then generate the complex surfaces and export into a mesh or objects and hence to a BIM system.

BIM systems (including SONATA, REfLEX and Revit) were and are notoriously bad at generating complex, curved surfaces. They have even been blamed for ugly architecture because of the difficulties of generating interesting surface shapes. It can be done but is less than straightforward, Rhino and Grasshopper enable the user to move away from these restrictions without getting involved in the heavy maths that might have otherwise been required.

BIM OBJECTS

A 'BIM Object' is a model of a component (e.g. tile, chair, sink), or assembled components (e.g. wall, roof, switchboard), or a space containing components and assemblies (e.g. a modular/pod kitchen or bathroom). BIM Objects can be generic (often referred to as requirement objects comprising geometry and object specification) or supplier specific (e.g. a precise representation of a manufacturer's component/assembly or space design).

It is important that as many stakeholders as possible are involved in creating data in a format that can be integrated into the model. This wider supply chain is now encompassing manufacturers and suppliers that create BIM Objects that can be downloaded directly into the Building Information Model, thereby simplifying and speeding up the process of creating detailed model data. This also provides benefit to the manufacturers/suppliers as its products are more accessible to its client base. The manufacturer/supplier can also be aware of who is accessing its products.

BIM Object can be used at any stage of the design and construction process. If the BIM Objects are actual model files they can be very data heavy and can significantly increase the size of the overall model. In such cases it is common to use the generic or object files, provided as part of a particular design software package, in the early design process and replace them with actual supplier files, later in the process. If, however, the BIM Objects are accessed via a cloud system, object data can be kept separate to geometry and accessed as and when required, via a GUID (Global Unique Identifier). In such cases the actual supplier files are much smaller and can be used from the early design phase onwards.

Usually the BIM Object is created and owned by the manufacturer/supplier of the particular object – it is responsible for the geometry and data that sits within the object. In the case of BIM Objects created as space designs, the owner may also be the developer responsible for the space design (e.g. government estates for, say, schools or prisons).

The common way to create an object is to design it in proprietary software and present it as a file. This process can be time consuming if multiple proprietary software formats are required as it necessitates object creation in each software. However, this problem is being addressed in, for example, the Cloud hosting service provided by BIMobject® (www.bimobject.com). It has a unique solution to this problem in that it provides free technology called BIMscript® which, working in combination with the modelling and optimisation features of Rhino, enables fast BIM content delivery in multiple file formats, providing one integrated process that generates all industry formats in one go and user friendly parametric objects for Revit, SketchUp and ArchiCAD.

QUESTIONS FOR VENDORS

There are some questions that might be asked of the system vendors.

It is worth remembering that BIM is a team process, where actions and models are shared across the teams, and so the system should reflect this. Sharing a building model requires a carefully designed approach that involves servers and crafted software to parallel "real-time" access for an entire team. This is crucial to the whole process.

Ask the vendor how many people can work on the model at the same time. The vendor needs to specify the system configuration required for your situation — questions such as Internet requirements, servers as well as the ideal computers. It might also be worth considering a growth path.

Different computers, level of detail, specific design requirements, number of users and most fundamentally, the software used will have a natural limit on project size. Only global BIM solution providers handle "localisation". When selecting your BIM tool, be aware of how the vendor's solution deals with local design standards. These vary from country to country and state to state. It might include local project and listing templates, and model sets following local rules.

There are a number of different sources of BIM Objects. Not all work on all systems. Ask what are available and what the vendor provides, relevant to your work. ArchiCAD has some 2000 predefined components built-in. They are easy to find and are well set out within the system, meaning they are readily available, fit the BIM standard and so on.

Most system providers have some components provided with many more available on the Internet provided by several organisations (BIM Object and UK NBS) across a range of disciplines. Drawing standards and so on may or may not be met by components loaded from the different suppliers, but still, it is probably better to go for the larger libraries maintained externally.

Are the BIM objects available up to date and do they reflect what you need; finished ready to order bathroom fixtures or generic steel bars?

Each user has functionality specific to their needs. Ask the vendor about your particular needs. A significant difference is that Revit has different versions for different professions, in particular, installation engineers, construction engineers, mechanical electrical and plumbing engineers, structural engineers and installation engineers, let alone architects.

ArchiCAD has limited ability to manipulate structures, requiring plugins. Although in theory, everything can be done in both, specific platforms referencing the same database is a preferred route.

Do not count line styles or how long something takes (unless it is significant) to compute, nor should the decision be based on which system has more or better functions, unless of course, the system has a feature that is essential to your work. Again what the stakeholders are using and your experience are probably the main factors. There are more Revit systems than anything else by quite a margin, threatening that at some stage, provided Autodesk continue to develop it, Revit will become dominant if not already.

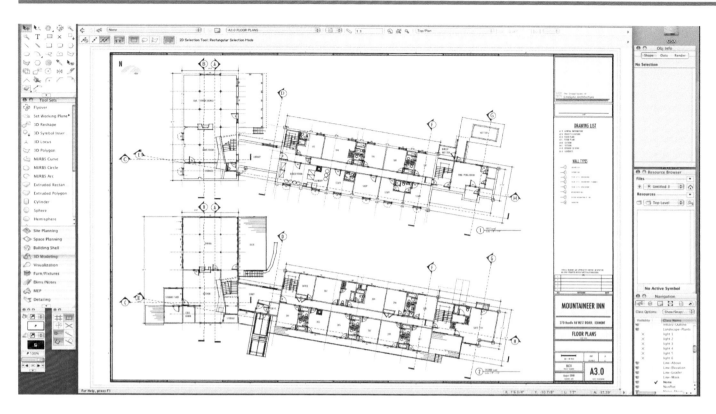

Figure 42. *VectorWorks Screenshot © Vectorworks*

It should be noted that some of the other systems are either weak BIM or not BIM. They say redefine BIM to make their systems apparently BIM compliant, so beware. The fundamental test is that whenever the model changes, and it will change all of the time, every piece of information updates automatically without running this program or pushing that function. Ask the vendor to move something in plan and establish if it has moved in elevation and the 3D, the bill of quantities and everything else you are interested in. All information must be up to date all of the time. Full BIM should enforce this.

Every design office has a variety of computers. The question is what machines have sufficient computational power and storage to house and run the BIM job. BIM is computationally intensive.

Having skilled designers waiting for the machine to perform tasks can be very expensive and frustrating, so a powerful computer available is preferable. If fast, it will be still loved in years to come. Sometimes the cost of the machines is seen as significant by some AEC users. In the end, other costs such as training, developing expertise, restructuring, building specific tools and frameworks

all tend to overwhelm this figure. The other issue is that software grows and slows. The software vendors build the latest versions of their software to run on the latest hardware. This hardware gets faster and faster and larger and larger. If you update your software, it will soon become slow on the older technology. New software expands to fill the space and cycles available on the latest machines. Again buy the fastest, largest machine possible.

The user or the user-base is a factor in equipment choice: give the user what he wants to use. If the user is uncomfortable or unhappy, then watch out. It is a bit like buying a car. Let them have the machine of their choice.

Inadequate workstations, incompatible equipment, poor software selection, insufficient bandwidth have been cited as reasons for hardware write down. BIM places significant demands on systems and users; the user issues can be addressed with proper training, the system by choice of software and powerful hardware.

BIM functionality must be distributed between all users while centralising the model. Each user must have the largest display, the fastest machines, and fast, seamless connectivity in order to maximise their contribution. It is however, not sufficient to have a simple hub as optimising the network is crucial to the integration of BIM into the firm's culture.

Depending on the geographical distribution of the firm and of the other companies sharing the model, a local area network or fast Internet may be needed. This should allow access to resources, files and software. The technology needs to be able to bring the data together, with everyone's efforts being saved to the central location with the BIM objects and model being updated appropriately. Copying data from within the model to send to stakeholders or users breaks the tightness of the BIM guarantee that everything is up to date.

Portability can also be an issue. Being able to take the computer on site, the next job, on the plane, to meetings, for presentations and just for the ability to do some work at random times, on the train, over breakfast; the laptop is king, mobile BIM rules. The issue with that has been the limited screen size. An HDMI connection to a large permanent high-resolution screen on the desk is ideal. One must have the fastest processor, a good amount of memory and preferably a fast disk drive, preferably SSD (Solid State Drive, it doesn't spin). This will all ensure that the machine has the longest possible life. Old machines are usually thrown away because they are too slow. Memory issues are usually addressable, though memory speed is also important. It's worth asking if the memory can be upgraded or extended. Rather than mention numbers here, which quickly go out of date, take the amount of memory the software manufacturer recommends and double it. Preferably make it a fast memory that can be added to later if necessary.

Some mechanism needs to be considered for backup. This is best made automatic as people forget to do it. The rule of thumb for backups is imagine losing a day's work, or a week or several weeks. The amount of time that causes significant pain sets the period of backups. The Author backs up to disk on the hour, to a separate file daily and to the cloud weekly. It is important to take the data offsite as if there is a theft or a fire, then recovering data is not an issue. The usual loss of data comes from human error. Save older copies and copies of occasional older copies. One needs to store historical data, as well.

The easiest way to do this is via a cloud system. Memory is cheap and providing you have a fast Internet link then a regular backup should not be an issue. A suggestion might be to make individuals responsible for their own backups but also have a manager back everything up. If the system is set up properly it will all just happen.

As stated, get the fastest processor and 64 bit. It is better not to go to overclocked processors. It's a bit like buying a highly tuned car, the chances of it failing are significantly greater. Unfortunately, one cannot compare different processors looking at just their clock speed. Different operating systems and how the code has been written can also have an effect. The best way to compare processors is by looking at a benchmark test for each processor. This is usually some computationally intensive process, somewhat similar to the BIM problem, that is run to completion on each machine and is timed. The latest computers have several processors, each processor with multiple cores. What this means is that there are a number of usually identical processors in the computer that can work independently, producing an overall result faster. That is the theory. If there are different processes or people using the machine, then it will be faster. If there is just the one program running, then it will depend if that program has been multi-threaded. All modern Intel machines have multiple cores and many have dual processors.

Most programs are not multi-threaded and the few that are expect little in improvement. Revit has released a multi-threaded version for wall-clean-up, hidden line removal and print functions. Multi-threading can lose performance.

Dual processors have a greater effect and are generally recommended in a BIM environment. Many BIM rendering engines are optimised for several processors so speed improvements in the rendering are significant.

A company called "vbim" manages the cloud platform for Revit and ArchiCAD users. They provide firewalls, servers, static IP's and all associated hardware and expertise to set up this aspect of the system very quickly. It claims to be scalable and resolves the issues of hiring IT staff. They also have a cloud based GPU system to provide fast image generation.

BIM DIRECTIVES AND LEGISLATION

Many Governments around the world have legislated in the use of BIM or advised on its use:

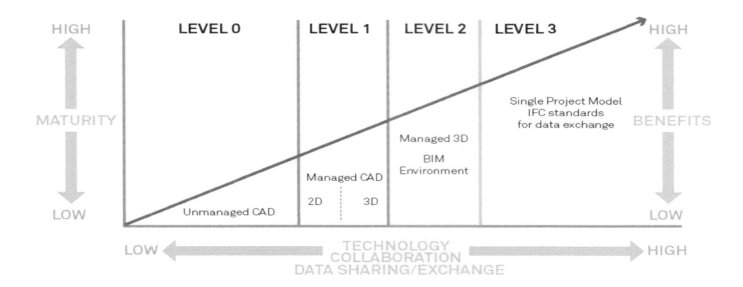

Figure 43. *Stages of BIM Implementation vs Benefits and Maturity*

- Australia's "Framework for the Adoption of Project Team Information and Building Information Modelling"[3] giving BIM guidance and preferred adoption methods. Although there are still no blanket mandates in Australia, a number of steps are being taken to implement BIM across Australia. For example, the NSW Government have mandated the use of BIM for the Sydney Metro Northwest[4]. (Some of the earliest users of BIM were from Perth and Melbourne, as outlined in earlier chapters. The reason for this is that the Author has Australian roots and was a frequent visitor during and after the development of SONATA and REfLEX.)

- EU directives encourage public procurement to use BIM, especially with the announcement that on the 26 Feb 2015 that the French and German Governments were also supporting BIM in various ways. The Germans in particular are mandating the use of BIM by the end end of 2020[5]. France has mandated the use of BIM in April 2017 as part of the strategy of digitizing the CI. In Italy the "BIM Decree" provides for the mandatory application of BIM from 2019.

---

3 https://www.acif.com.au/resources/strategic-forum-for-building-and-construction/a-framework-for-the-adoption-of-project-team-integration-and-building-information-modelling

4 "Future of BIM in Australia" Morrissey law publication https://morrisseylaw.com.au/future-of-bim-in-australia/

5 http://www.bimplus.co.uk/news/german-government-set-mandate-bim-transport-projec/

- Scandinavia led the world with the first BIM legislation in 2007.

- In the USA, BIM is fairly widely used though not all states in the USA have legislated for it. Wisconsin was the first in 2010 with all projects over $5million being required to use BIM. The US had an adoption rate of over 70% by 2014, but with 50 different states, there is no single BIM initiative and no mandatory legislation. An update on this is that the U.S. General Services Administration has mandated that new construction must use BIM. It specifically referred to the use of the 4D, 5D and 6D technologies.

- In China, the Ministry of Housing and Urban-Rural Development has outlined the use of BIM in "industrialization, urbanization and agricultural modernization in its 12th Five-Year Plan". It is not, however, mandatory to use BIM. In Hong Kong, the Institute of Building Information Modelling (HKIBIM) the road-map for BIM was formulated in 2014.

- In Russia, the Minister of Construction, Housing and Utilities has announced that the use of BIM is "compulsory" for all construction projects commissioned by the Russian Government from the start of 2019.

- The UK leads the world, not in terms of adoption, but rather in terms of standardization with explicit legislation in place.

The UK Government has implemented a four year program to implement BIM and currently requires collaborative 3D BIM on its projects, to reach (see figure 43) Level 3 BIM, targeted for 2019. To quote the 2016 Government budget *"The government will develop the next digital standard for the construction sector – Building Information Modelling 3 – to save owners of built assets billions of pounds a year in unnecessary costs, and maintain the UK's global leadership in digital construction."*[6] All public projects over £50m must use BIM to level 2 and by 2019, use level 3. A more realistic view is that we are now at the start of a ten-year program to implement level 3.

In the UK the degree of implementation is described by Levels. Level 0 is defined as unmanaged CAD and Level 1 is CAD in 2D and 3D.

In general terms, Level 2 BIM involves "collaborative working" using a BIM model but not necessarily sharing a single model between different parties. The information is passed between different users' models. According to Graham Jones at the CIC this means, *"building models must include parametric data attributed to a 3D environment that is managed using proprietary formats and bespoke middleware."*

Level 3 represents full collaboration between all disciplines by means of using a single, shared project cloud-based model. All parties can access and modify that same model, and the benefit

is that it removes the final layer of risk for conflicting information. In March 2016, The UK government reasserted its commitment to Level 3 BIM as part of the Budget and so the UK will be moving in this direction for years to come. This mandate has been set as one measure to help in fulfilling their target of lowering costs by 33%, faster delivery by 50%, lower emissions by 50% and improving exports by 50%. These are ambitious, but some of them are possible.

In the U.S. the overall guidelines give a direction for the future. The National BIM Standard-United States provides a standard aimed *"at the standards needed to foster innovation in processes and infrastructure so that end-users throughout all facets of the industry can efficiently access the information needed to create and operate optimised facilities."*

NBIMS-US™ V3 provides construction industry professional guidance in the use of BIM. This document was developed by open consensus originally based on a 2007 document by 30 experts. It is built around a set of International (ISO) standards including Information, Conformance, Normative and ISO standards. A layer of technical publications are based around these standards and deployment resources around that.

It is aimed at software developers and vendors, and at CI professionals who *"design, engineer, construct, own and operate the built environment"*. These are developed by buildingSmart.[7] These documents and the standard are still work in progress. It is possible to belong and contribute to this group.

---

6   Professional Construction Strategies Group 2019
    https://pcsg.co.uk/2016/03/17/hm-government-budget-2016-
    commits-to-development-of-building-information-modelling-3/

7   https://www.nationalbimstandard.org/

Implementing BIM 5

We have seen from the earlier chapter that BIM comes with great promises, but with the decision to "Let's do it" come many questions. BIM is not just about technology but, perhaps as importantly, it is about new work practices and a whole new vocabulary. It comes with a range of challenges, steps, considerations and pitfalls. Among other things, legalities and the steps inherent in building the models, need to be examined.

BIM is not just a technical tool. The tool is important but just as important, perhaps more so, is that it involves a new methodology, a new way of working. BIM implementation is not only about choosing which system to use: it is in fact the integration of technology and process. Various designers can work on the shared model at the same time, where technology and computer software allow them to see and share data interactively. This sharing of data within the model integrates information naturally, almost as though everyone is in the same room.

We have seen that BIM is best not done alone: it works optimally as a collaborative effort. At its conception, few organizations using it actually worked together, solitary BIM was rife, and case studies show the first attempts at joint working.

The MacLeamy Curve[1] (Figure 44) shows how BIM affects the design process in terms of effort and cost. Curve 4 indicates a significant design effort early in the process, during the schematic design and design development stages. Because the virtual model is built during this stage, considerable effort is required here. From the same figure, the cost of making changes is much lower during this phase, hence leading to a lower-cost design. It also gives the

designer a greater opportunity to optimize and change the design for a much lower cost.

Conversely, traditional design methods lead to higher cost, less ability to change the design and hence, less optimization. Building a virtual model means that design can be optimized and costs lowered.

Constructability of a project requires direct interaction between those who design and those who construct. Architects and engineers can use the virtual model to determine details and construction sequence, ensuring that those on-site can investigate the model to see how it should be done. Being able to work through the model with the stakeholders enables all aspects of the building to be examined, as literally, all the data is available in different forms to everyone.

Another example is the integration of information between the scheduling and the logistics of the sequence of construction. Estimating costs also requires close liaison between vendors and stakeholders, so having a single source of shared data ensures agreement.

Sustainability also requires input from a broad range of sources, users, designers, builders, facilities managers and regulatory bodies. This is not a simple interchange of information but rather a complex interaction of ideas, designs and requirements. The analysis of sustainability is also best done early so that the model can be optimized for this.

Defining implementations steps are critical to the success of a new installation. In terms of basic management principles, the steps in the implementation process should be:

---

1   "Collaboration, Integrated Information, and the Project Lifecycle in Building Design and Construction Operations", Journal of Engineering, Project, and Production Management, 2012

THE MACLEAMY CURVE

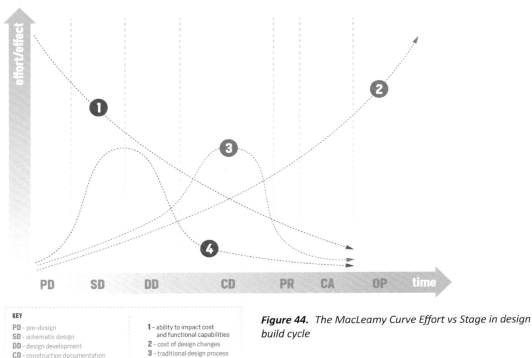

KEY

PD - pre-design
SD - schematic design
DD - design development
CD - construction documentation
PR - procurement
CA - construction administration
OP - operation

1 - ability to impact cost and functional capabilities
2 - cost of design changes
3 - traditional design process
4 - preferred design process

Figure 44. *The MacLeamy Curve Effort vs Stage in design build cycle*

- Select the BIM champion, someone with ability, sufficient power and determination, normally the architect
- Ensure the "champion" is fully supported at higher levels
- Determine which of the benefits you wish to achieve
- Ensure the complete design team is behind the new methods and is prepared to change their processes to ensure success, and, of course, that they have access to the BIM

These may seem obvious, but without each in place, success, together with its benefits, is more elusive.

Various changes will happen once these initial management steps are taken. The first and perhaps most important decision is the selection of the BIM software itself. The main systems or add ons are Autodesk Revit, Graphisoft ArchiCAD, Bentley Triforma, Tekla BIMsight, Navisworks, BIMobject, BIMx, Trimble, Procore, SketchUp, Allplan, Autodesk BIM360 and others. All are conceptually similar though some have a better grasp on BIM principles. Some are easier to use, some deal with larger projects, some are more 3D than BIM. These have been discussed in the previous chapter.

Traditional approaches to buying software work: talk to the vendors, try a few systems, talk to some BIM consultants. You will soon have a feel for what suits you. Unless you're building individual houses, buy a powerful machine: BIM needs power. Ensure a fast internet connection to allow sharing of files and then it is time to have a go.

Make sure you have an idea of what you are trying to achieve, where you want to see the changes; understand that it will take time to see these changes, progress along the learning curve is rarely smooth. There is a large body of help sitting a few clicks away. Consider joining the SMART alliance[2] where you can network with global BIM leaders.

BARRIERS TO BIM

With change comes resistance, most commonly fuelled by fear of the unknown. Changing from a conventional drawing environment to BIM brings several challenges and risks. Various studies have

---

2   www.buildingSMARTalliance.org

been performed to analyse the barriers of implementing and running an BIM-based environment. [3] [4]

Employing experts with relevant experience has been identified as an issue with BIM. As a rapidly expanding technology, there are never enough experts to fill the positions available. As the technology matures and the rate of growth slows, then one would expect this situation to ease.

Unfortunately, the actual work process of creating a model, controlling input, monitoring access, creating protocols and leadership can be daunting. BIM managers are very important in the process, especially those with expertise in what is becoming a less niche field.

Another barrier has been identified as the acquisition of any new technology and the direct costs associated with it. Software, hardware, and potentially training are all costs that must be incurred before there is any return.

Potential legal issues regarding model ownership and responsibility have been an issue, though with specifically tailored contracts for these situations becoming available, this situation has eased.

Change in work practices seems to meet internal resistance in companies, though it's usually solved by management and education.

Another issue can be that if the client has not requested BIM, is there any need to volunteer? This is a changing situation with legislation and the realisation of the potential benefits.

It is recognised that first time working through and completing a BIM project, from design to construction, is probably slower, it may have unknowns and is a learning process. Finding a suitable project can be problematic, but, in the end, BIM has been and is successful because the benefits significantly outweigh the problems. Careful management of the process can ease all these issues.

BUILDING DELIVERY METHOD AND BIM

BIM encourages collaborative, integrated working methods: with all designers working closely on a project. Integrating teamwork, building early virtual models and BIM into the requirements of any delivery method can be challenging. The American Institute of Architects recommends a new *"project delivery approach that integrates people, systems, business structures and practices into a process that collaboratively harnesses the talents and insights of all participants to optimize project results, increase value to the owner, reduce waste, and maximize efficiency through all phases of design, fabrication, and construction"*.[5]

Design-Bid-Build is still the most common building delivery method for both public and private buildings. The client chooses an architect to create a schematic design. Structural and building service engineers are then involved to design related structures and services, developing it into a final design. The resulting drawings are used for the construction bidding phase.

Although not the generally preferred method for use with BIM, this can still make use of many aspects of the BIM process, especially as it offers security to all bidders regarding the true extent of the works required. This is based on the assumption that the model is available to all. This is still an issue in many countries, especially where designers want to restrict the use of their ideas and documents.

BIM processes, however, are invaluable to another popular construction procurement process, Design-Build, where architect, engineer and contractor work together from the start, collaborating to produce buildings on time and budget for more prudent clients. For these clients, who often go on to manage their own buildings as well, the opportunities of live Facilities Management BIM files are icing on the cake!

BUSINESS MODELS

New business models are needed to work with BIM. This includes the new close working relationship between stakeholders in a project as they affect roles, risk, timing, and reward.

One of the most significant changes resulting from BIM is the design process itself, where designers now assemble the virtual building and construction information is produced automatically

---

3   "Potential Barriers to Implementing BIM in the German AEC market", Both, P., 2012, http://papers.cumincad.org/

4   "Barriers to implementing BIM in the Construction Industry from the design perspective", CTW Chan Journal of System and Management Sciences, 2014

5   "Improving Building Industry Results", https://damassets.autodesk.net/content/dam/autodesk/files/bim_and_ipd_whitepaper.pdf

from it. This virtual model, often constructed by different people, different teams and different disciplines, needs a manager. The manager is responsible for its maintenance, for the participants following correct procedures, probably being "shown", and the model being secure and backed up.

The business structure that surrounds BIM can vary from lesser adjustments, based around trust, of existing contracts to the full Project Alliance, pioneered in Australia. The latter typically bonds the owner-builder-designer team around a single agreement that ties all to a non-claim, total project success-based reward system, sharing the value of the success delivered. This is a strong model for sophisticated clients and experienced, trusted, design and build partners. The risk and reward in this model is value-based, not fee for service-based, where all parties share in both the up and down sides.

New business models should be applied to the industry to achieve sustainability. Without incentives and penalties, there is little incentive to change. BIM offers the opportunity to relate all costs and earnings directly to the long term sustainability of the project, in terms of performance delivered to customer and society.

With the BIM benefits of an accurate prediction of performance and cost, some business incentive should be passed to those responsible. Conversely, bad performance should be penalized. This is fundamentally changing or extending the cost-plus business models to a performance-based model. This subject is discussed further in Chapter 15.

BIM itself has "forced" practitioners to work together in design and build coordination. Similarly it could be used to force a change of the business models. BIM additions for value configuration, cost structure, partners networks, and whole life-cycle model evaluation.

INTEGRATED PROJECT DELIVERY

Integrated Project Delivery (IPD) is a business model for designing and constructing buildings that integrate people, systems, business structures and practice. Utilizing BIM, it aims to optimize value for the building owner, reducing waste and improving efficiency through the different phases of the building construction. It achieves this by engaging, at an early stage, all parties concerned with the project in decision making integrating design and construction decisions. IPD teams are formed early, close to project inception, diverse in terms of disciplines involved in the design process.

IPD provides continuity of coordinated information, eliminating errors, optimizing design solutions and reducing communications errors. It allows models to be managed for the optimization of all needs from start to finish. Through these early design considerations and by providing the opportunity to optimize the design, the incorporation of sustainable strategies is more effective.

IPD teams can decide how the BIM is developed, accessed and used, and how the information is shared. Software decisions are based on experience, functionality and interoperability. The level of detail, tolerances, component sets, and perhaps the purpose of the model, are all clarified in the process.

In more detail, further decisions are made around the model, especially to what extent is it to be used for costing, the functionality individuals expect and how the BIM acts as part of the contract document. The relationship between the BIM and the other contract documents should be established to ensure a cohesive legal approach to the contract process.

The methods for hosting, maintaining, managing and archiving the model must be determined as part of the process. Compensation for work on the model and eventual ownership must be decided between the parties, where project quality, schedule and sustainability all come under the IPD umbrella, directly related to the use of BIM.

BIM PROCUREMENT

Accurate quantities and programme forecasts come directly from the BIM. Suppliers, delivery times and delivery arrangements can and are, all built into it. BIM has a very positive effect on procurement in terms of quantities, timing and source.

BIM can be used for Just-In-Time (JIT) procurement[6], delivering the right material at the right moment. Inefficiencies are known to occur because of stock shortages, incorrect orders size, inadequate storage facilities, perishable materials at the wrong time (especially concrete), multiple handling of materials and out-of-sequence deliveries. The advantages of JIT in construction include reduced inventories, smooth flow of materials, improved

---

6   "Just-in-time purchasing", Ansari & Modarress, Collier Macmillan, 1990

quality and increased productivity, all of which have been well recorded. [7] [8]

BIM improves the benefits of JIT. Visualization helps clients and other stakeholders understand complex projects. Site layout storage and handling of materials in the model solve the issues of materials handling. Improved certainty and predictability also assist in ensuring uninterrupted work-flow. Integration of JIT with the time model sequence also helps with scheduling, storage, and reducing congestion on-site. In addition, waste can be reduced; reduced waiting times, transportation and handling, waste processing and inventory. Other advantages that BIM brings to JIT are better flexibility, enhanced space usage and improved quality.

Through a shared mode , BIM also assists with information dispersal about deliveries, machinery maintenance, sequencing, accurate material quantities and potentially physical limitations of delivery equipment within the structure being assembled.

APPLICATIONS OF BIM

BIM has broad-reaching applications across a project well beyond straight design. Many of these have been discussed elsewhere in this book, but for completeness, an attempt is made to list many of the applications (adapted from [9]):

- **Design & Engineering**
- Project Definition
- Conceptual Design
- Architectural Design
- Engineering Design and Analysis

- **Town Planning**
- Project Management
- 4D Scheduling
- 5D costing
- 6D time sequencing
- Value Engineering

- **Collaboration**
- Information Integration
- Information Distribution
- Document Management

- **Bidding**
- Quantity Surveying
- Scope Definition

- **Construction Planning**
- Shop Drawing
- Fabrication
- Surveying
- Radar scanning
- Field positioning
- RFI Management

- **Building and Facilities Management**
- Maintenance
- Monitoring

- **Risk assessment**
- Metrics and Impact Scheduling
- Conflict Identification Resource Assessment Visualization Surveying Analysis Feasibility
- Site Management Integrated Project delivery
- Decision Knowledge/Data aggregation
- Accelerated decision making
- Impact Exposure
- Accountability and auditability

- **Project Assurance**
- Quality Control
- Predictability
- Delivery Optimization
- Cost and Risk avoidance
- Information Control

- **Finance**
- Co-ordinating funding with execution

INTER OPERABILITY AND FOUNDATION CLASSES

Interoperability is the ability for information from different systems and vendors to be passed between one system and the other. In the past, this has meant loading drawing files between the various systems. Now the aim is to load complete models (though in some respects the model is never complete). Usually, the design, construct and management process requires software from different vendors. When data is passed between different systems, a rigid protocol or format needs to be established depending upon the way the communication happens.

The common method of sharing designs is by specifying a format for information that the different systems can all read. Any

---

7   "Towards Construction IT", Researchgate.net, Ballard & Howell, 1995

8   "Just-in-Time Management of Precast Concrete Components", Journal of Construction Engineering and Management, Pheng & Chuan, 2001

9   "BIM implementation strategies", Ashcraft and Shelden, Gehry Technologies, Hanson Bridgett

systems compliant with the process can write and/or read the agreed format, for which there is any number of formats, including the Industry Foundation Class (IFC). As an example of sharing files, IFC is used to submit a Revit concept design to the Integrated Carbon Information Model (ICIM) website, where another user can examine and reconfigure it as required.

Originally conceived as the International Alliance for Interoperability in 1994, the IFC has become the dominant means of transporting BIM data. As an aside, in 1995, the Author and Bob Wakelam were invited to demonstrate SONATA technology and join the IAI (founded 1994). Both felt, perhaps incorrectly in retrospect, that Autodesk had an overwhelming influence in the group. The fact that our direction was so different was also an issue.

Since then the IFC specification has been developed and maintained by building SMART. Essentially the IFC defines several different file formats, some readable in editable form, that system manufacturers use to "transport" information between the various systems. These formats define how different sorts of objects relate to each other.

IFC uses architecturally meaningful objects to describe real-world building items. These objects contain data associated with the architectural object.

IFC allows:
- Cross-discipline coordination of BIM
- Data sharing and exchange across IFC-compliant applications
- Re-use of data for analysis

Most vendors, including Revit, Vectorworks and ArchiCAD subscribe to IFCs allowing both export and import of IFC format files.

Building simulation tools commonly use IFCs allowing the relatively straightforward adoption of building geometry and other data. The use of IFCs with these systems leads to low-cost virtual buildings, shorter simulation times and better use of the simulation.

Unfortunately, there are various issues in sharing standard data formats, including IFCs. These are not reasons not to use them, but in using the IFCs, be aware of the following:

- If data is copied then you will have duplicate data, at some point you will more than likely have a data clash. Also note:
- Only basic data is transferred, not the complex parametric definitions nor links to other objects.

- References to external files and related data are immediately lost, while at the same time giving many users the impression they have the latest information.
- The use of Standard data formats such as IFC tend to cater for the lowest common denominator among the various users of the format. High level logic or connections can be lost.
- When the BIM is modified, inevitably during design development, and data is saved again and again into the standard format, then the work loading it into all the external programs must be redone.

Having said all this there is a time IFC has its place. There are many other applications where extending the views in this way would be useful, discussed in later chapters.

PROJECT INFORMATION MODELLING

Project Information Model (PIM) is the "information model developed during the design and construction phase of a project." PIMs consist of an assembly of distinct models, generated by the different disciplines to create a single model of the building, including non-graphical data and associated documentation. The PIM model evolves from the Design Intent Model initially into the Virtual Construction Model.

PIMs build in richness as the project progresses, until handover, where the complete data set is passed to the asset's owner or end-user, potentially via COBie. The PIM becomes the Asset Information Model or AIM once a project is complete.

Through the Design Intent Model, design suppliers should show architectural and engineering intentions. This might include massing diagrams, simple symbols representing generic elements of the design and critical parts being shown in more detail. It might also include:

- Spatial and Architectural form
- Outline services and structural design
- Outline specifications
- Existing utilities
- Existing surveys
- Access details
- Outline infrastructure requirements
- Basic shadow and shade analysis
- Demonstrate compliance with the client brief
- Outline site and landscape design
- Local visual impact
- Basic construction and phasing information
- Basic cost plan
- Possibly construction plan

The Virtual Construction Model provides increasingly detailed design information that might include:

- A detailed construction sequence
- Links to project management tools
- Safety and access information
- Temporary work information, scaffolding, retention structures, etc.
- Safety briefing information
- Crane zones
- General site information

The Virtual Construction Model may become the full BIM model effectively.

Once the construction is complete, the PIM has been compared against the constructed building, it will become the AIM.

CONSTRUCTION ORIENTATED BUILDING EXCHANGE (COBIE)

COBie exists to automate the job of the FM at handover of the completed building. In the past, a huge effort has been required to create, review, validate and transcribe hundreds of documents relevant to the new building into Operations and Maintenance Manuals.

COBie was created in 2007 to provide operations, maintenance and asset management electronically. It is a data format for the publication of a subset of building model information focused on delivering building product information, rather than geometry. COBie is closely associated with BIM and is typically exchanged using XML. These formats support some of BIM interoperability, and many BIM systems interface with them.

COBie is a specification for facility asset information delivery, including two types of assets: equipment and spaces.

COBie helps the project team organize electronic submissions approved during design and construction, subsequently delivering consolidated electronic O&M manuals with little or no additional effort. This data can then be loaded directly into FM or asset-management software.

Several different formats of COBie are available: the full version and the lite version. The full version uses text format (STEP Based on IFC) that can be loaded directly into spreadsheets, for which templates are found on the COBie website. Exchanges of small bits

of COBie data about individual assets is another type of exchange supported by COBie Lite.

COBie data, from the designer's point of view, is the set of all the schedules found on the design drawings. From the contractor's point of view, COBie data is simply a way to produce construction submissions so that information is repeatedly copied and re-organised.

BUSINESS VALUE OF BIM

BIM brings many advantages to the process of lifetime design, maintenance, operation and demolition of buildings.

While these benefits are sometimes obvious, various studies have been made to quantify the degree of improvements in bottom-line case studies. There are also residual benefits of using BIM, which make a significant difference to the business case but are less easy to measure, discussed later in this section.

In total, BIM costs typically account for less than 2% of the overall net value or the overall net revenue. The initial investment in implementing BIM includes acquisition of software, hardware and training, but are spread over all subsequent buildings. Some of the variable costs associated with a new BIM project include the level of detail of the model, the complexity of the project and the experience of the design team. Unfortunately, contracting the model out to be built by third parties can be a significant additional expense.

In general terms, Young[10] showed that the return on using BIM was positive for 70% of contractors and owners, while around half of engineers and architects reported a positive ROI and only 7% indicated a loss. The average perception of the ROI on BIM is between 11% and 30%, with one third suggesting ROI of around 1000%, roughly 10 times the cost. Azehar suggest an ROI on BIM of between 140% and 39900%.[11]

The main tangible savings are:

- Better communication because of 3D models (79%)

10 "The Business Value of BIM: Getting Building Information to the Bottom Line", Young, N.W., Jones, S.A., Bernstein, H.M., & Gudgel, J.E., McGraw-Hill Construction Smart Report, 2009

11 " Building Information Modelling (BIM) Benefits, Risks and Challenges" Azhar, S. Hein, M., and Sketo, B., McWhorter School of Building Science: Auburn University, AL. US, 2009

- Better at winning projects (66%)
- Improved outcome, fewer change orders and RFIs (80%)
- Energy Savings in improved building design
- Improved cost and estimates
- Improved building outcome through easier consideration of alternatives
- Maintenance scheduling

One of the most significant figures claimed was the reduction in change orders or in reworking by as much as 90% by using BIM[12].

Energy saving is also of particular importance in this age of climate concerns. BIM allows energy conscious decisions and testing to be made from an early stage.

Costing and budget estimating are the process of predicting costs of materials, labour and time, for which BIM provides a guarantee in the design, build and operation processes. Automated quantity take-off is a natural by product of BIM and will soon allow the general contractor to perform a quantity take-off in a few minutes. However, BIM models need to be constructed with sufficient information in order to extract this information.

LEGAL ISSUES ARISING FROM BIM

There are two fundamental legal questions that arise using BIM: ownership of the model and its implications, and issues arising from the collaboration. Sharing a model means that the various contractual arrangements will differ from traditional contract models.

However, the built virtual model has a value beyond the design phase in maintenance, management and monitoring of the actual building. Each BIM is made of different parts or sub-models assembled by the appropriate parties, where each party owns their particular role and are responsible for maintaining that part.

Model security can be assured technically and can be protected from external change and viewed as "read-only" by other parties separately. The building owner or another interested party may contract to receive the model from the "appropriate parties". Ideally, a copy should be sent to a "library" as part of a national registry.

---

12 "Building the Future", Khemlani, L., 2006, BIM Symposium at Uni. of Minnesota, AECbytes, URL: http://www.aecbytes.com/buildingthefuture/2006/BIM_Symposium.html

In the UK, with the current level 2 BIM, few significant changes to contracts are required. For example, it might include a schedule of services for each designer to reflect their responsibilities with regard to BIM protocols and the BIM model. The ownership of the BIM model and the responsibility for maintenance and hosting the model is shared. The UK Construction Industry Council publishes a BIM Protocol designed for use with Level 2. The working group went on to suggest Level 3 will require action to be taken to ease the early adoption of integrated working.

The RIBA has produced a BIM Overlay to the RIBA Outline Plan of Work while the AIA has developed standard forms to assist parties with integrated working concepts when working with BIM.

Insurance is affected directly by the use of BIM. Using BIM to construct a virtual building in advance of the actual construction reduces the risks of errors, omissions, clash and rework. Only recently have insurance products been developed that reflect project sharing, reduction in risks, and that align with the goals of the integrated project.

INTELLECTUAL PROPERTY RIGHTS (IPR)

Copyright normally vests with the author of a work, except, when they are an employee, but where the work is collaborative, the copyright is shared. Copyright usually vests with the author rather than the party that commissioned the work and gives the owner the exclusive right to authorise or prohibit the exploitation of the work.

Where the author of the created work is employed, employment contracts normally stipulate that this work would belong to the employer. Thus a developer of HVAC wanting to use the BIM model may need a license or assignment of copyright from the owner. Joint ownership would mean shared Construction Industry Council (CIC) BIM protocols referred to in the previous section, which specifically deal with ownership and licenses of the BIM model and other electronic media. It supersedes any conflicting Intellectual Property (IP) provisions in an existing contract as they relate to BIM.

Under the Protocol, companies have an obligation to license any material they use that is owned by third parties, where the organisation in charge of a building project would represent that it will procure those rights. Companies that create value in the BIM may suffer reduced control of the IP under the Protocol.

When the BIM model is created for a particular project, under the Protocol it is licensed to the employer on a royalty-free basis. They may sub-license the material to other companies, perhaps limited to those involved in the project. BIM Protocol is a straightforward document with a clear mechanism for incorporation into Agreements, but in the U S. the IP value of a BIM model is handled differently.

In the USA, the BIM model ownership is specified by contractual terms. These terms should consider the format, ownership and future uses of the BIM model. The issues that arise from this include liability of the model creators. It is prudent to expressly exclude implied terms and determine bespoke warrant obligations as to the use and performance of the BIM model.

MULTI-DIMENSION BIM

BIM is used to assist in scheduling and in the management of the construction process. Extra dimensions are added to BIM to allow for cost and time in the BIM process.

4D BIM relates to scheduling and programme information defining when a particular object should be required. It might include lead time, time to become operational, installation time and the sequence in which components should be installed. Just-in-time (JIT) delivery of materials to the site becomes possible with this information. It helps with planning work, ensuring safety, and achieves a vision of the construction sequence. Chapter 11 deals with Construction Management with some of the original work in 4D and 5D BIM.

5D BIM relates to the estimating and cost aspects of the building. This might include capital costs, running costs and the cost of replacement or maintenance. 5D BIM enables the ability to see the costs in 3D form, be notified when changes are made to model affecting cost, and assuming the presence of 4D programme data, track predicted and actual cost over the project. The costs associated with the objects might depend on quantities (for example, a window gets bigger the number of bricks reduces) and non-modelled quantities such as temporary works. It might also reflect the construction quantities rather than the design quantities.

These two terms, 4D and 5D, are at odds with the original naming convention first used in the 1990s. The Taylor Woodrow brochure, reproduced in Figure 145, refers to 5D modelling as scheduling of the project, and 4D referred to costs.

6D BIM includes information to support FM and operation. It also covers the sustainability targets for a building. This could include information on the manufacturer of a component, its installation date and required maintenance. Information such as energy use, thermal properties and carbon cost allows some energy analysis

to be performed together with details of how the item should be configured and operated for optimal performance, energy performance. Lifespan and decommissioning data would also be included.

LEVEL OF DEVELOPMENT (US)

The Level of Detail or Development is the amount of information contained within an object making up the building model. As the design progresses, detail about the design of the individual elements will increase. For instance, in Figure 45, the system knows there is a wall, but the actual makeup of the wall may not be decided till later in the process. The state of where each object is at is called the Level of Development.[13]

The UK and the USA have a different approach to the Level of Development and Level of Detail correspondingly. The Americans use a Level of Development approach that allows a tight hold on the exactness of information in the model. This works well but has been expanded to take up all aspects of the BIM model. The Level of Development of a BIM model increases as the project progresses, moving from a rough approximation to detail which is known to be accurate, suitable for construction and possibly building management. The American Institute of Architects called it the Level of Development rather than the Level of Detail as it refers to the decisiveness of the information not the amount of information though there may be some correlation between the two. The US-LOD is defined as how seriously you take the data associated with the model. One must be careful as a highly detailed object may be placed at level 100. LOD 100 says the detail is not reliable.

A first-stage model may comprise of a single block representing the form of the building relative to those around. This would be a LOD of 100 and would be suitable only for basic measurements such as volume or area.

With a LOD of 200, the actual dimensions of the beam are correct, and the concrete is specified. At LOD 300 the item is much more detailed and accurate, and LOD 400 catches every last detail. It is accurate on all counts and reflects what has been supplied.

The BIM Forum has released a reference document detailing the levels of detail of BIM elements at different stages in the design and construction process. This standard defines, on an element by element basis, a wide range of information and attributes, some with a LOD flag. For instance, with an assigned LOD for each parameter, an open web steel joist has a type, a

---

13   "Level of Detail", U.S. General Services Administration, gsa.gov

designation, a length, a depth, a weight per length, a range of loading parameters, fireproofing parameters, dates, exact joist dimensions, and so on, to a total of 71 parameters.

"These definitions allow model authors to define what their models can be relied on for, and allow downstream users to clearly understand the usability and the limitations of models they are receiving. The intent of this Specification is to help explain the LOD framework and standardize its use so that it becomes more useful as a communication tool."[14]

CIC BIM PROTOCOL (UK)

The "Construction Industry's Council BIM Protocol: Standard protocol for use in projects using Building Information Modelling" can be used in projects to define obligations, liabilities and define limitations on the agreed use of a model. The BIM protocol was defined as part of the CIC's response to the UK Government BIM strategy.

More precisely, it defines the contractual obligations between Employer and Contract Team, setting out the rights, identifies responsibilities for particular tasks and can require compliance with different standards.

It can be included directly into a contract with an incorporation clause making easy use. It can be downloaded from the Construction Industry Council website[15].

EMPLOYERS INFORMATION REQUIREMENTS (UK)

Employer's information requirements (EIR) define the information that will be required by the employer to develop and operate the completed built asset. It forms part of the tender documents on a BIM Project.

Typically it defines the model, including the Level of Detail (LOD), in three main areas, Technical, Management, and Commercial.

- Technical defines software used and Level of Detail of the model
- Management defines roles, responsibilities, work planning, collaboration processes and model review meetings
- Commercial defines BIM project deliverables, the timing of data drops and BIM competency assessment

The BIM Task Group provide a "Core Content and Notes Guide" and "Employer's Information Requirements" on their website[16]. This is a specimen document to be included with tender documents for the procurement of a design team and construction. The National Building Standards (NBS) also provides a tool kit to assist in the preparation of these documents.[17]

Suppliers respond to the employer's information requirements with a plan from which their proposed approach, capability and capacity can be evaluated. By the end of the project, it defines the information to support the maintenance and operation of projects systems and components.

BIM EXECUTION PLAN (UK)

BIM Execution Plan (BEP) is a document shared and agreed by all parties in the project team. It specifies how they will work together through the BIM process to deliver the requirements of the Employment Information Requests. BIM Execution Plans typically detail team roles and responsibilities, deliverables and time scales associated with them, approval procedures and logistics, formats and conventions for file sharing.

The BIM Execution Plan is in three parts, the pre Contract, the post Contract and the Master Information Delivery Plan. The pre Contract plan would include project milestones, goals for information modelling and a plan setting out the capabilities and experience of the bidder. The Master Information Delivery Plan, submitted by the successful supplier, sets out when, by whom and what project information is to be provided.

Plans can be found on the CPIC (Construction Project Information Committee) website.[18]

HEALTH AND SAFETY

Shorter construction times are the first natural safety by-product; less time on site is less time for an injury to happen. Linked to this is the greater use of pre-assembled or prefabricated units which would mean less craftwork work on site. Reduced change orders also simplify the work at hand again shortening the work process.

---

14 "Level of Development Specification", BIM Forum, Oct 30 2015

15 http://cic.org.uk/

16 "Core Content and Notes Guide", BIM Task Group, www.bimtaskgroup.org

17 https://toolkit.thenbs.com/articles/employers-information-requirements/

18 https://www.cpic.org.uk/

LEVEL OF DETAIL

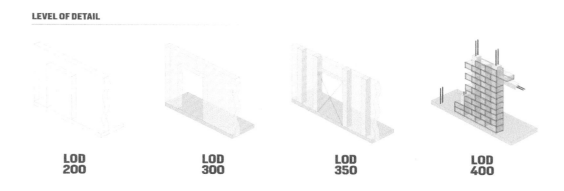

| LOD 200 | LOD 300 | LOD 350 | LOD 400 |

Figure 45. *A Masonry wall element progression from LOD 200 through 400. After BIM-Forum*

Planning using the model can avoid hazards of different types; vehicle or crane traffic drop zones, fall zones, electrical issues, and so on. In difficult situations, the model could be used for virtual on-site training, taking personnel through different potential hazards. Safety training creates a higher level of vigilance hence reducing possibilities of accidents.

Modelled use of hazard signs would remove the casual nature of the current layout and obscured or badly placed signs can be avoided.

The use of sensors to detect problems in the structure linked to a BIM model would be helpful. In the Engineering Case studies, the use of sensors in the coffer-dam wall, linked to objects in the model, gave notice of any movement in those walls.

Selection of temporary works around a structure can be modelled with safety in mind. As the construction progresses, partially built frames and systems can be tested in terms of strength and safety.

Should an accident happen, the model and its features can be examined to determine if adequate measures were considered in the BIM.

The presence of hazardous materials, radioactive or bio-hazards can be modelled and mapped for training. In extreme cases, immersive training with Virtual Reality may be used to ensure familiarity with the issues. Operating objects discussed in later chapters will give further realism and an indication of the dynamic issues involved.

Early Case Studies

BIM demonstrates application across a wide range of disciplines. The next seven chapters show various projects that benefited from BIM. These projects date from 1985 until fairly recently.

They were done by forward-thinking architects, engineers and technicians of the time. Being few, these pioneers had made a journey of discovery, finding benefits and uses which are now commonplace. With the groundswell of BIM use, particularly with sharing the workspace, it's hard to recall that finding and convincing these early users to change their work practices was sometimes difficult. They provided a testbed for many different aspects of BIM, trying out the ideas and finding work practices that suited them and the system.

The Author worked with many of these professionals, understanding their requirements and adapting software to match. Examples included realistic (for the time) lighting algorithms, construction management time phasing, alternative tunnel shapes, road alignment algorithms, cut and fill models, and multi-user access to the database.

No suitable commercial imaging programs were available at that time, so considerable effort was spent in adding accurate shadows, transparent effects, textures, and so on. As a point of reference, the first coloured Macs appeared several years later in 1988.

These studies show the work of architects, engineers and plant designers, local government, power, mechanical, electrical and industries. These projects bring a comparative, historical perspective to modern BIM.

They should bring an understanding and perspective of BIM together with an appreciation of its history, its development, its structure, and its power.

Architecture

6

Architects started using BIM in the mid-1980s. The first practices to go out on that limb (it was then a long limb) were GMW Partnership (now Scott Brownrigg) and PRP Architects in the UK and Peddle Thorp Melbourne in Australia. The revealed benefits for these users were coordination of plan, section and elevation drawings and 3D image rendering exemplified in the National Tennis Centre (Figures 63 through 65) and Barclays Bank (Figure 84). Many of the other advantages were lost because they were sole practitioners.

The Author attempted to make the generated images as acceptable as possible, within the speed constraints of the machines of the time. The word "acceptable" is used in that true realism was not achieved for several decades after this time. Because of this, other tools were provided to make the images more attractive. The contemporary image Figure 217, an image from a live scene, is almost indistinguishable from its actual photo. Almost all the images in this section are from algorithms written by the Author.

Other features that architects enjoyed at that time were the coordinated windows and drawing types, parametric elements and automatic cavity closing and hole generation from windows and doors in walls. Being able to change the design and parameters in one view, and immediately see the changes in all the others was regarded by some as almost magical. As the user base extended, multiple users sharing a database became important.

Various problems arose for these early users; the libraries of the different components were limited to some hundreds of items, though tools were provided for them to generate bespoke or specific elements. Other issues were the low resolution of the screens and the slow speed of creating videos.

BIM brought the ability to create and communicate design intent, especially for architects. Not only were the buildings

represented but also how the building fits into the surroundings. These were of course of considerable interest to the client, planners and neighbours. Shadow envelopes of buildings, both new and existing, can be crucial to any new development and its surroundings. Block models have been used with great effect to get an idea at different times of the day and of the year, see Figures 92, 93 and 97.

Society increasingly requires architects to design buildings that are ecologically sensible and sustainable. Virtual buildings gave and give BIM users the ability to review sustainability in terms of thermal design and materials used.

One of the critical issues outlined in Chapter 1 is that the different professions tend to design and duplicate design within their ivory towers. BIM, as we have discovered, encourages the various professionals to share their design space. Virtual models allow architects and engineers to collaborate more easily so that clashes between services and the different parts of the structure can be detected and resolved, and errors and omissions consequently reduced. Even if they don't work together on the same model, checking everything through the computer is bound to improve things. Pre-constructing the virtual building may mean that more effort is spent in the early part of the design phase, in order to avoid changes later in the construction phase when changes cost more, but it may not. Building the virtual model is often faster than drawing alternatives, especially for repeat projects where templates can be used, which of course leads to fewer change orders, lower costs and a faster build.

Some of the benefits to the architect in constructing a virtual building are less obvious. Seeing how to build it and how it will look brings a confidence to the whole design team in that they know it will work, not only in terms of a building and its usage but also in terms of how easily it can be constructed. Simply as a consequence of everything having been considered beforehand,

Figure 46. *REfLEX Image Melbourne Sports & Aquatic Centre by Jeff Findlay 1998 © Peddle Thorp Melbourne*

Figure 47. *REfLEX Image Melbourne Sports & Aquatic Centre by Jeff Findlay 1998 © Peddle Thorp Melbourne*

Figure 48. REfLEX Images Exhibition Street Apartments Melbourne by Jeff Findlay 1998 © Peddle Thorp Melbourne

Figure 49. SONATA Image National Tennis Stadium by Jeff Findlay 1987 © Peddle Thorp Melbourne

Figure 50. *SONATA Image residential and leisure complex, from Sonata RUN magazine*
Courtesy Sutherland Craig © SONATA Systems 1989

Figure 51. *SONATA Image Exhibition Centre Melbourne by Jeff Findlay 1989 © The Building Modelling Company*

Figure 52. *RεfLEX Image ESSO HQ Southbank Melbourne by Jeff Findlay 1997 © Peddle Thorp Melbourne*

Figure 53. *SONATA Image St Patrick's Church, Melbourne Cityscape 1988 by Jeff Findlay © The Building Modelling Company*

Figure 54. *St Patrick's Cathedral, Melbourne, SONATA Image Courtesy by Jeff Findlay 1998 © The Building Modelling Company*

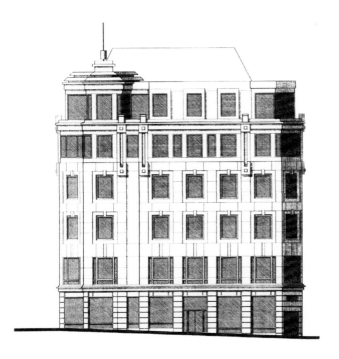

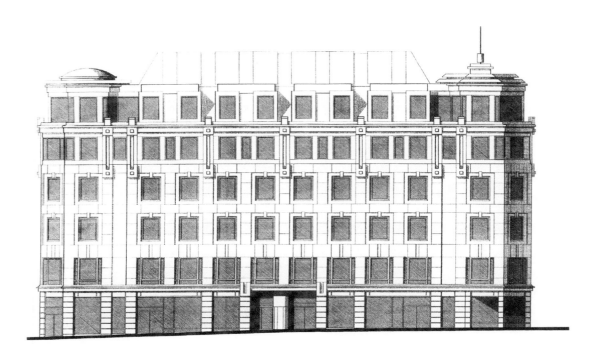

Figure 55. *SƆNATA Elevations Proposed Office Development City of London 1991 by Seifert Architects*

Figure 56. Coordinated SONATA Drawings from a single model Carlton Gate, Maida Vale, 1987 © PRP

Figure 57. Photos as built Carlton Gate, Maida Vale © PRP

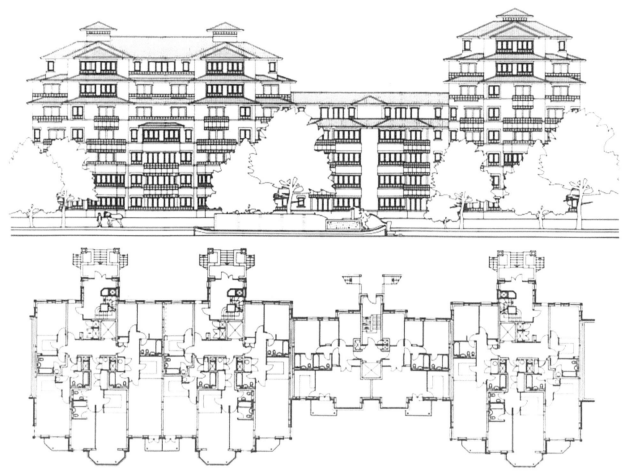

Figure 58. *Coordinated SONATA Drawings from a single model, Carlton Gate, Maida Vale, 1987 © PRP*

the client will convince stakeholders more easily and will improve control over of the construction of the building with time and cost savings.

CASE STUDY: CARLTON GATE, PRP ARCHITECTS

An established practice PRP designed Carlton Gate, Maida Vale, London in 1986/1987. Initially, PRP used the drafting system RUCAPS, moving to SONATA in 1987, because of the parametric objects and 3D. PRP was one of the first architectural practices to use Building Modelling and move to BIM.

Carlton Gate is a development of approximately 600 apartments, including accommodation for nurses and other health service employees together with a Health Centre. Low towers run along the Grand Union Canal, while a wall of buildings shelters the site

from the Harrow Road, with small scale courtyards creating a close-knit urban environment.

Until the mid-1980s the Carlton Gate site had been occupied by old buildings used for health care. The South West Thames Hospital Board put the site out to a staged competition to maximise site value. 3D Building Modelling used by PRP to win the competition looked at massing options and early design ideas while monitoring changing densities and sales value.

One of the results achieved was to build separate blocks which could be sold during the construction phase, allowing construction to be sped up or slowed down consistent with demand. The construction of reinforced concrete, using flat slabs and columns, is a form of construction pioneered in the UK at Chelsea Harbour which became widely used as the late 1980s development boom gathered pace around the world.

Figure 59. SONATA Images Lloyds Bank by Richard Rogers © Giuliano Zampi 1986

CASE STUDY: LLOYDS BUILDING, RICHARD ROGERS ARCHITECTS

Buildings with elaborate interiors (and exteriors in this case!) and sophisticated lighting need to be modelled to allow the architect to verify the effectiveness of their design, to optimise the environment and to enable the client to see and easily comprehend precisely what they are getting.

Giuliano Zampi used SONATA to model the iconic Richard Rogers designed, Lloyd's Insurance building at 1 Lime Street, London. The building was innovative in that the services, including staircases, lifts, conduits and water pipes, are on the outside of the building, leaving an uncluttered open space inside. There are three main towers and three service towers around a central space. At its core is the large Underwriting Room, which houses the Lutine Bell. The Underwriting Room is overlooked by galleries, forming a 65 metre, naturally lit, high atrium through a vaulted glass roof. Galleries open onto the atrium, connected by escalators through the middle

Figure 60. *SONATA Image The Peak Tower, Sir Terry Farrell, 1993 © Giuliano Zampi*

Figure 61. SONATA Image The Peak Tower, Sir Terry Farrell, 1993 © Giuliano Zampi

Figure 62. Photo Peak Tower Hong Kong Sir Terrace Farrell

of the structure. Higher floors are glassed in and are reached via the exterior lifts.

CASE STUDY: THE PEAK TOWER, FARRELL ARCHITECTS

In 1993, Peak Tower Hong Kong underwent a HK$500 million redevelopment into a new entertainment and retail space. Sir Terrance Farrell's design is bold and beautiful, shaped to catch luck following local tradition. In addition to retail shops and a sightseeing gallery which was completed in 1997, the building contains a Mini Motion Theatre and Ripley's Odditorium.

Guiliano Zampi's convincing portrayal of Farrell's iconic design helped win the design contract. Accurate lighting and modelling, together with Zampi's dramatic eye, contributed to producing a series of strong, yet representative images. Figures 60 and 61 show external studies, which for the time, were a likeness to the photo shown in Figure 62.

CASE STUDY: NATIONAL TENNIS STADIUM PEDDLE THORP ARCHITECTS, MELBOURNE

Peddle Thorp, arguably Australia's leading architect for many decades, were amongst the first in the world to embrace Building Information Model technology. Principally used as a modelling tool generating external studies, Peddle Thorp used many features such as costing, coordination, massing and environmental impact studies with shadows, and so on. They are prolific designers, so examples of their BIM work can be found throughout this book.

Soon after its launch in January 1988, the iconic National Tennis Centre, subsequently renamed the Rod Laver Stadium, was designed using the first BIM system, SONATA. The brief from the NTC Trust was to provide the State of Victoria with a world-class complex that could operate as the Grand Slam tennis venue and a multi-use entertainment centre for the rest of the year. Peddle Thorp was engaged to design the multi-purpose venue, and the architect Peter Brook was also commissioned to help with

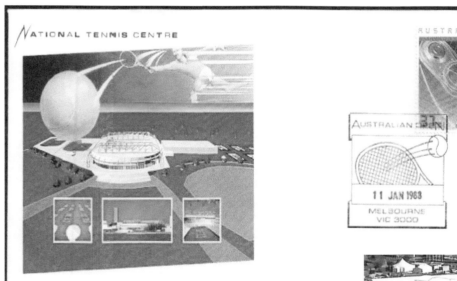

Figure 63. SONATA Image on an Australian First Day Cover stamped 11 January 1988 by Jeff Findlay

Figure 64. Photo National Tennis Centre as built

Figure 65. *SONATA Image National Tennis Stadium by Jeff Findlay 1987 © Peddle Thorp Melbourne*

design and project management. After intensive discussions with management, Peddle Thorp came up with the idea of a retractable roof, which had not featured in the brief nor anywhere else in the world at that time.

Original designs by Peddle Thorp in 1982 had been discarded owing to prohibitive costs, and resurrected when it became more than a tennis centre. As a multi-use entertainment complex, it could feature major concerts throughout the year, almost commonplace in major sporting venues throughout the world these days.

"This allowed the centre to be economically [viable], and from its base as a sports and entertainment centre, a garden square and function centre were added in 1990". For the accompanying Hisense Arena in the same sports precinct, the central idea of seating which could be raised to allow it to become a velodrome was developed with Graeme Samuels and Peddle Thorp. "After suggesting the idea of the raisable seats for the Velodrome, we were given five weeks to make it work," Peter Brook, Design Director of Peddle Thorp said.

The multi-purpose arena was completed in 2000 at the cost of $65 million. The venue has a fully retractable roof taking 20 minutes to close; dynamic seating configurations that are raisable, retractable and removable; larger floor area than the original Rod Laver Arena, including corporate suites and function rooms.

CASE STUDY SOLE ARCHITECT AND SMALL PRACTICE: PETER DEW AND ASSOCIATES

In a recent interview with the Author, Peter Dew gave his view of small practices and Building Object Modelling (as BIM was known then).

"Many architects trained in the early 1970s were aware of the potential for Computer Aided Design but reluctant to swap their drawing and presentation skills, often resulting from many years of training, for AutoCAD type drafting systems, really nothing more than computerised drawing by numbers. Their vision of CAD was much more closely allied with what BIM has become."

Peter Dew RIBA was one of them, passing up on SONATA owing to cost, not only of the modelling software but computer workstations of the time, and REfLEX, more economical but hardly attractively priced for a small practice. The change came when the then top visual computer company, Silicon Graphics Inc, produced a small workstation, the O2 in 1996, and PTC bought out the Author's REfLEX software, re-releasing it as ProReflex in 1997. The SGI O2 was Video ready with integrated software, bundled with Adobe Illustrator and Photoshop, both of which could be utilised alongside ProReflex to run the office from the one machine. Soft-Windows allowed standard Microsoft programs to be run as and when necessary.

"ProReflex software was around the same price as AutoCAD at the time, though there were an increasing number of alternative drafting systems, including BricsCAD, EasyCAD, ZwickyCAD, etc., that later brought the price down. However, it was the first

Figure 66. ProReflex Image Possibly First Stables in BIM 1997 ©Peter Dew

Figure 67. Adobe Photoshop and ProReflex collage 2000 © Peter Dew

Figure 68. *Reflex Image 1997*

Building Object Modelling software priced at a level that made it feasible for a sole practitioner and sometime small integrated architect and engineering practise to consider computerising. Whereas most AutoCAD users were still employing draftsmen, BOM [BIM] users could quite easily do it all by themselves, especially with a little training, and telephone and on-line help. The internet was still new, and file transfer by email and ftp a god-send, the fax machine being the only competitor.

Looking back, it took a major leap of faith, but after a few months it was possible to produce all the Design and Tender drawings for simple buildings alone, including 3D Views and walk-throughs, completely unaided by assistants and draftsmen, and within a year or so, to do the same thing for relatively large projects, apartment buildings, hotels and the like.

PDA produced fully coordinated drawings, including both the Architectural and Structural sets, with current 3D views on

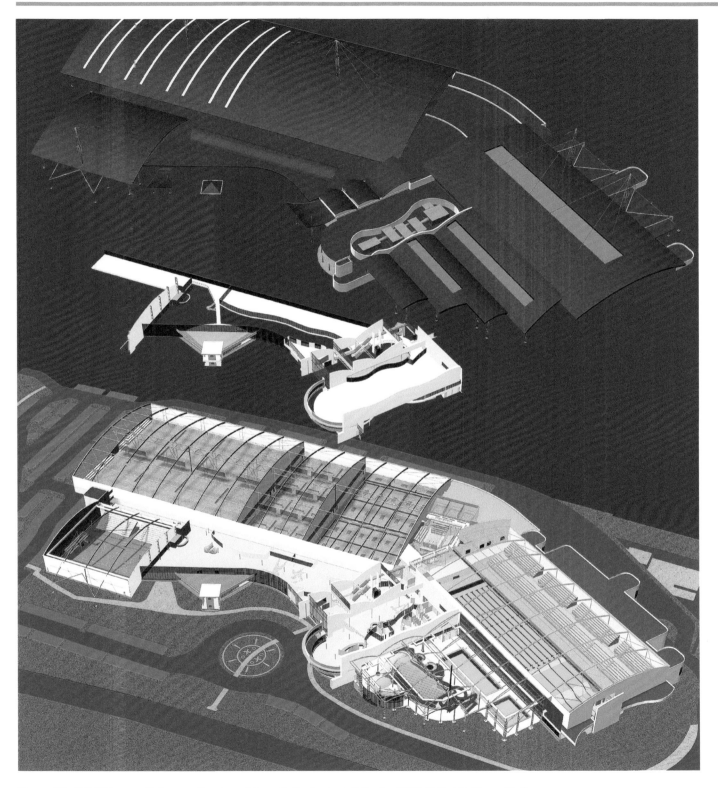

Figure 69. REfLEX Image Melbourne Sports and Aquatic Centre by Jeff Findlay 1997 © Peddle Thorp Melbourne

the cover sheets, cut and pasted building owner's photos into presentations, gave a helicopter fly around the local international airport to confirm the location of new buildings, which to knock down and redevelop, even mundane things like revised branding colour schemes, and prepared Bills of Quantity from the Schedules, sure they were right, all more economically than going to separate Quantity Surveyors.

However, where it was necessary to collaborate, exporting drawings in file formats compatible with other CAD systems was possible, as it remains today with Revit, but the advantages of coordination were immediately lost."[1]

ARCHITECTURAL ANIMATION

Animation of live projects was becoming more common towards the end of the 1980s, and by the 1990s, numerous examples exist

---

1   Interview with Peter Dew 2018

from various systems. Camera paths, viewpoints, lighting, and so on, are all defined.

This technology has changed hugely in the last twenty years mainly as a result of initially the movie industry and later the development of computer games engines. The later Chapters of The Future and Retail Information Modelling display some of these images.

The images here show some of the earliest architectural animations. They were all done as part of clients wanting to show their stakeholders exactly what was to be built.

Teesdale Development Corporation (TDC) built a commercial complex on the River Tees as part of a major urban regeneration scheme (Figure 72). Designs for the various buildings and the river barrage were drawn up, and a full-colour fly-through was made, a seven-minute video featuring a drive across the barrage and a walk-through of the buildings were completed and viewed to great

Figure 70. *Which is real? SONATA Image & Photo Australian Broadcasting Corporation by Jeff Findlay 1988 © Peddle Thorp Melbourne*

Figure 71. *REfLEX Video Frames from Gatwick 5D refurbishment video 1996 © Taylor Woodrow*

Figure 72. *SONATA video for Teesside Development Corporation by Mark Edwards © CADAIM*

Figure 73. *SONATA video for Tesco PLC for supermarket approval in Croydon by Mark Edwards© CADAIM*

acclaim in 1997. It is generally recognised that video sequences of buildings and structures are a compelling way to reveal the look and feel.

Figure 71 is from the Gatwick refurbishment video as is discussed in later chapters. Figures 55, 70, 74 and others show a number of different drawings and images from SONATA and REfLEX. Both classical and modern buildings are represented, where it will be noted that older buildings, in particular Figure 55, a hidden-line drawing, give the feeling of etched drawings, typical of the era of the buildings themselves. Adding shadows into this type of drawing, as in the City of London elevations by Seifert Architects, lends a feel of relief to the building facade. A more modern

building, but beautiful image is Figure 88. Taking this to the extreme, several render packages were developed to produce images comparable with water-colours. One of these was Picassa from Cambridge University.

The modern Melbourne Sports and Aquatic Centre by Peddle Thorp (Figure 69) shows shiny metallic surfaces for an ultra-modern representation of the stadium.

By contrast, Figure 74, the drawing of the University of Birmingham Music Hall is a hidden line cutaway of a modern building.

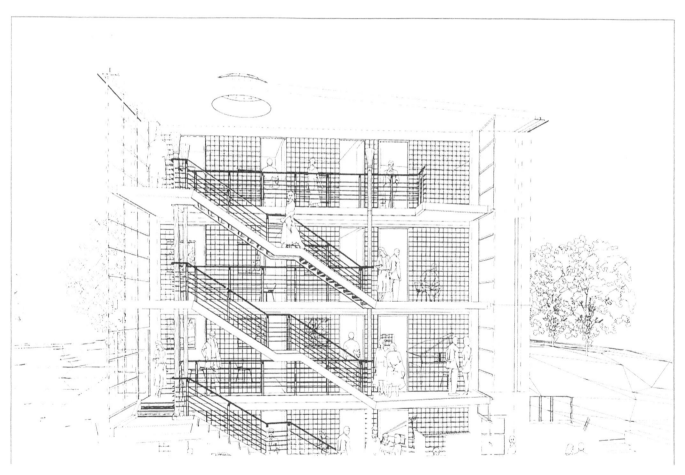

Figure 74. *SONATA drawing, University of Birmingham Music Practice Rooms 1994 © SONATA Systems Inc*

Figure 75. *SONATA Image by Mark Edwards © CADAIM*

Lighting & Shadow Studies

7

One of the natural spin-offs of BIM is the complete 3D model from which images can be rendered. These rendered images have been one of the most utilized parts of BIM models. Here we introduce the subtleties of rendering, how the clients have, in the past, used different lighting, shadows, surfaces and effects to achieve images that are attractive and representative.

Seeing the virtual building and how it fits in with its surroundings, the style, colour, texture and appearance, are the most important issue for most of us. Visualization of the construction workflow, site offices, buildability, even the power used for different lighting effects are important features needed by everyone else, from engineers to interior designers. Light and shade play an essential part in terms of comfort, happiness, efficiency and general sustainability: however, they are not quantifiable in layman's terms, hence an image is often the only solution for presentation. This drove a pioneering technological spirit in the Architecture and Computer industries and for many years led to the title architect being misappropriated for computer engineers and software designers.

Once a model is set up, images can be regenerated at any time including after any design changes and the results easily compared. Rendering of models is somewhere between a science and an art. Steady progress in quality has been made since the first images were made in the 1970s. The first usable colour computer monitors only became available in the early 1980s, and the first digital cameras were only available for sale in the late 1980s. Colour printers of sufficient quality weren't available until the mid-1990s, and even then, were exorbitantly expensive.

Various tools were used in these early systems, typically:

- Multiple light sources time and date based
- Shadows
- Smooth shading

- Textures
- Transparencies
- Fog effects
- Realistic coloured spotlights
- Metallic surfaces
- Camera facing "flat" figures
- Material surfaces
- Background images (sky etc.)
- Ambient lighting effects
- Dynamic colour changing within the image

The code for each effect had to be written, modelling what was happening mathematically. These are all taken for granted now with special hardware to apply each of these features. These basic effects produced a quality sufficient to give the end-users an idea of what they were going to see.

Giuliano Zampi and Conrad Lloyd Morgan wrote "Virtual Architecture"[1] that displayed many of these and other images using SONATA.

SURFACES AND LIGHTING

The perception of space relates to the surfaces and how light falls on the surfaces that limit that space. The light in a scene falls on surfaces where some colours are absorbed, some diffusely and specularly reflected, and some internally reflected (water). This light then comes to the eye, perhaps reflected from yet other surfaces. Semi-transparent or highly reflective surfaces add their own complexity. The natural and artificial light of different mixtures of colours interacts to produce a very complex collage of colour for every single point on every surface.

---

1   "Virtual Architecture", Zampi and Morgan B.T., Batsford, London, 1995

Figure 76. *SONATA Colour Changing within the rendered image*

In a camera, this light is taken through receptors that interpret the colour and add their own characteristics. Film cameras add a whole new level of complexity and possibilities.

This collage massages the eye to form an image of the environment. Being able to view these surfaces in a way that conveys the architectural and interior design intent is an important

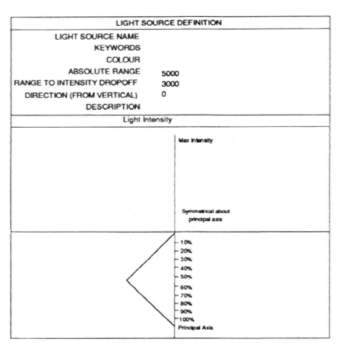

Figure 77. *SONATA User defined spotlight with variable angular intensity. © SONATA Systems Inc 1988*

capability. Add texture to surfaces, mirrors, transparency, roughness, fog, shadows (umbra and penumbra) and possibly motion and one realizes how complex the interplay between light, surfaces and the eye is. The eye has evolved over millions of years to distinguish between subtleties in images, so is a well trained, well-adapted tool for seeing the world.

How a building is lit, both inside and out is critical in how it is used, how people perceive it, and how efficient and sustainable it might be. It can convey diverse feelings from austerity to inviting, from power to common. The BIM model, together with good rendering software gives the designer a potent tool, where he can find an attractive, efficient solution to the building or structure.

In rendering an image, every specific effect has to be mathematically calculated. Appendix 1 discusses this in some detail, but the essence is as follows:

The point, or more accurately the area, of the scene behind each pixel on your screen determines the colour of the pixel. The machine does not "know" what surface is visible, and what is visible is the first calculation. Once the contributing surfaces of the area are found, colour, roughness/smoothness, shininess, the texture, the shadows and so on must all be summed to determine the colour to be displayed. Of course, if the surface is semi-transparent, then this same process must happen for the surface that can be seen beyond the transparent one. The direction of the light, its colour, whether daytime or night-time, the intensity of light, and whether in shadow or partial shadow affect the different surface characteristics and the display colour.

BIM makes all of these possibilities available to test alternatives and try things out, to see how attractive or functional the results

Figure 78. *SONATA Image of Giuliano Zampi (right) working out virtually © Giuliano Zampi 1989, Image shows smooth shading, transparency, shadows, textures, spotlights and simulated reflections*

Figure 79. *REfLEX Image by Jeff Findlay © Peddle Thorp Melbourne*

are. However, modelling of these effects in software is complex and requires some skill to produce realistic, engaging images. Daylight externally, daylight internally, internal day and night-time artificial lighting, all require very different approaches in terms of definition and display. However, it is all extremely computer-intensive. Originally, rendering calculations took minutes and movies took hours or in some cases months. Rendering accurate and attractive images has become an industry in itself and all machines, even smartphones have hardware rendering capabilities.

In wanting to generate images that are realistic and faithful to the environment, extra detail must be considered. Daylight especially is affected by global position (longitude and latitude), atmospheric conditions, time of day and other factors. Shadows that result from daylight change continuously, becoming softer until diffuse at dusk. Similarly, artificial lighting needs to be modelled based on light colour or warmth, intensity, direction, fall-off, and so on; interior shadows are affected in the same way.

As suggested above, even transparent glass includes reflective surfaces, textures, colours, patterns and many other factors, affecting how light falls and is seen by the eye. Computer

simulated effects, with different degrees of success, produce an image of the scene, work out which bits of the 3D model is hidden from the viewer relatively simply, unless transparent surfaces or reflection are involved, comparing and determining the colour of each pixel in the view.

Figures 80 through 86 and 89, show buildings lit at night with different colours and different spread of intensity spotlights, but shadow bands on some images with spotlights were a bug in the early implementations of spotlights in SONATA. Named spotlights defined their radial spread of the light, light colour and intensity numerically, representing Lux (intensity/unit area) at 1m. Some BIM generated images are compared with photographs of the finished buildings, sometimes favourably even from over 25 years ago!

SHADOWS

Shadows are an important part of the perception of any building. Shadows are as important as the light which casts them. They exert immense effects on how a structure is seen and behaves with the environmental conditions. Shadows give the building form, the surfaces degrees of colour, brightness, texture and

Figure 80. SONATA Image with varied spotlight colours © Giuliano Zampi

Figure 81. SONATA Image BP Gas Station Modelled by Giuliano Zampi © Giuliano Zampi

Figure 82. *SONATA Image, architects Dixon Jones, Sainsbury, Plymouth © Giuliano Zampi*

Figure 83. *SONATA Gentlemen's Club Tokyo Image © Giuliano Zampi Architect Julian Bicknell*

Figure 84. *SONATA Image of Barclays Bank lighting study 1992 © GMW/Scott Brownrigg*

Figure 85. *SONATA Image Control Building Cardiff Bay Barrage Architect Alsop & Stormer courtesy © Giuliano Zampi 1990*
Winner of the SONATA Image of the year in 1990 awarded by t2 solutions
Entry in Royal Academy Summer Exhibition (10384) 1992

Figure 86. SONATA Image Peninsula Hotel Hong Kong © Giuliano Zampi

Figure 87. Photo of Finished building Peninsula Hotel Hong Kong © Giuliano Zampi

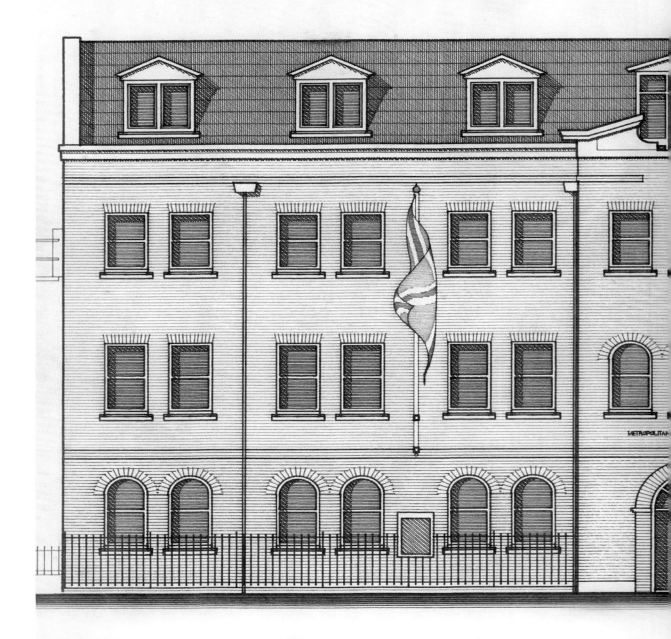

Figure 88. *SONATA Hidden line elevation with hatching and shadows, Metropolitan Police HQ 1990*
Under license from Metropolitan Police © Metropolitan Police UK

Figure 89. *SONATA image lighting analysis for RNT, London, Architects Staton St.John © Giuliano Zampi Architect*

Figure 90. *SONATA image of Golf Club Japan 1991 © Giuliano Zampi Architect Julian Bicknell*

Figure 91. *SONATA image of Golf Club Japan 1991 © Giuliano Zampi Architect Julian Bicknell.*
CAD image winner of 1991 UK CICA Gold Award

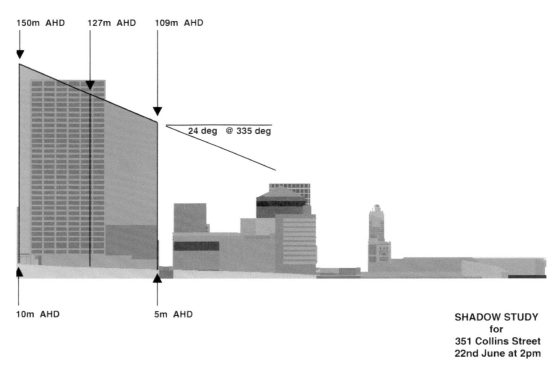

150m AHD   127m AHD   109m AHD

24 deg @ 335 deg

10m AHD   5m AHD

SHADOW STUDY
for
351 Collins Street
22nd June at 2pm

Figure 92. SONATA Shadow Studies by Jeff Findlay 1988/89 © Peddle Thorp Melbourne

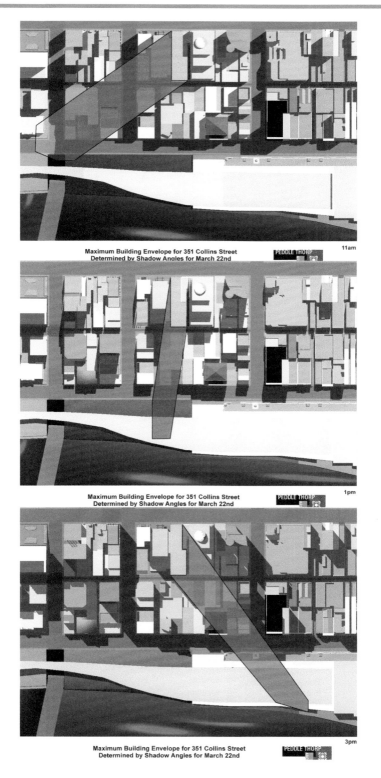

Figure 93. *SONATA Images Shadows for building envelope by Jeff Findlay 1988/89 © Peddle Thorp Melbourne*

Figure 94. SONATA Image Office Interior with spots and shadows by Jeff Findlay © Peddle Thorp Melbourne 1987

Figure 95. *SONATA Image Melbourne by night by Jeff Findlay © The Building Modelling Company 1987*

Figure 96. *SONATA Image Offshore Hotel Singapore © GMW/Scott Brownrigg*

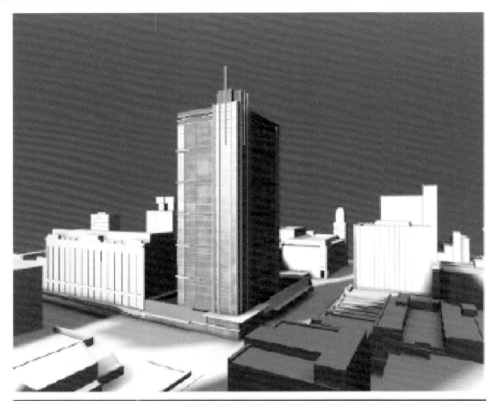

Figure 97. *REfLEX Images of CBD Melbourne by Jeff Findlay © The Building Modelling Company*

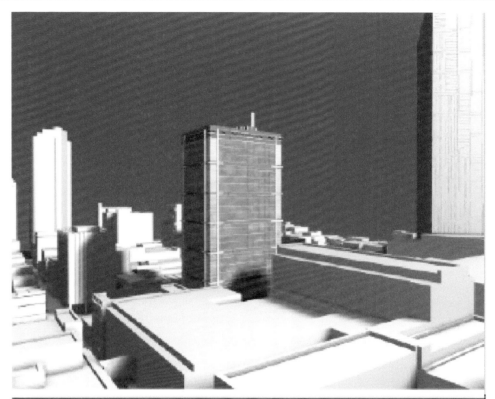

Figure 98. *REfLEX Images of CBD Melbourne by Jeff Findlay © The Building Modelling Company*

softness. When combined with transparent surfaces and objects, the shadows take on colour and complexity. Shadows provide respite from the sun and steal the sun depending on location and time, and to a degree, the people exposed.

Shadows are not black or white but come in changing colours and brightness levels. They are painted by the colour and intensity of the light, by the bright surfaces around them (ambient light and radiosity), they are broken by bright reflections and have penumbras that smudge them from nothing to black. Day and night, spotlight quality, and sunlight characteristics all will affect shadow, shade and surface perception.

Shadows are important to tie or hold objects together. A simpler image, Figure 245 has the Rietveld chair tied firmly to the surface by the shadows. Displace the shadow slightly, and the chair moves above the surface by a precise amount. Shadows can make a wall curved or flat, whereas in the view without shadows, the wall can be ambiguous. Certainly, a shadow can accentuate a shape, or its lack can make the shape amorphous.

Shade on surfaces is, according to Leonardo Da Vinci, a form of shadow. Where light is and is not in an image is crucial. How the surfaces are shaded, and we are not talking about how to draw or render the surface, gives great character or not to how the things appear. In order to see shadow, one needs light; it is the contrast between the two and how these are played that make for great or ordinary architecture.

Mechanically, the thermal, energy and light performance of a building is directly related to surfaces and shadows, not just on the building itself, but also on the surrounding buildings. These must be considered and are perhaps as important as the appearance of the form of the building. Designing a building for

different climates and for daytime and night-time is immensely complex. Arriving at an optimal, attractive and sustainable, complex modern building is even more difficult. We have seen that with a BIM model, we can at least move towards this ideal.

It is straightforward to set the sun position in the imaging algorithm automatically based on longitude and latitude of the site, where shadows would be generated for that setting according to the date and time of day selected. Peddle Thorp Architects used this functionality to generate a series of images of the Central Business District (CBD) Melbourne, Australia, to determine the impact of 381 Collins Street on the surrounding buildings for different times and days of the year; see Figure 93. Precise shadow envelopes are generated at specific times during the day throughout the year. The exact effect on the surrounding neighbours can be determined. Changing the light quality, perhaps from dawn to dusk, would change the nature of the shadows.

Night-time shots are more challenging in that they require explicit lighting points. These require colour or warmth, directional intensity, and whether they have hoods to cast particular shadows.

Building Modelling Company, Melbourne, built the whole Melbourne CBD in SONATA, allowing clients to check the visual impact of new buildings on their surroundings. St Patrick's Cathedral and the National Tennis Stadium images both contain views of their city model.

Early Retail Applications   8

This chapter, and Chapter 12 Retail Information Modelling, discuss the application of BIM to Retail. This chapter goes over the first uses and Retail Information Modelling is a current and future view of Retail and Information Modelling.

In the late 1980s, several well known Swedish, Australian and British high street shops, including Safeway, Coles, Asda and John Lewis, saw the benefits of having a single information model into which they could tie diverse information from around the store.

Store Modelling was, and remains a specialized application where parametric components are developed to represent the different fixtures, fittings and goods in the store. It is not simply speeding up refurbishment and refitting programs, but about linking the BIM of the store and its contents to related commercial databases. From 1986 SONATA parametric incorporated sales and product

profitability trends into the design process, leading to more cost effective layouts. The underlying database and the accompanying parametrics recorded quantities for cost-estimating, produced floor area calculations for correlation against Online EPOS data, and provided layouts based on accurate assessments of product profitability rather than using gut instinct alone.

Store design and fitting-out is an ideal application as layouts and designs for interior fittings, such as racks and gondolas, are easily replicated. Merchandisers used the system to assess the best position for product ranges within the store and several connected live EPOS data into the model, bringing retail information directly in from external databases. Reduced time and cost of refurbishments, optimized retail space, store quality, merchandise and range planning were improved. Schedules of operations and maintenance were achieved with several customers. Building Code compliance was also checked. Even the Bank of Turin, Italy connected their asset inventory in an Oracle database directly into the SONATA model to help with facility management.

Store Modelling System libraries enabled retailers to manage retail space by linking the comprehensive BIM models of the store to related commercial databases, thereby incorporating sales and product profitability trends into the design process. By creating a 3D prototype of the store, everyone involved in the decision-making process could see and understand the proposed design. Quoting John Stokdykl of the time: *"By walking through the model at customers' eye level, retail managers can work out where to place 'speeders' to attract customers into the next section, or where to put eye-catchers to slow them down."*[1]

Figure 100 shows an Asda retail store in the UK. Intelligent parametric aisles were created, and display carousels that automatically laid out the store for any "reasonable" shape of

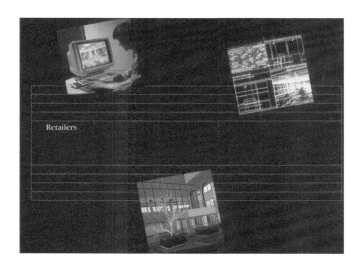

Figure 99. *SONATA Brochure Cover for Retail © SONATA Software Systems 1990*

---

1   "Run International CAD journal" Edition 40 published by "The Users Group" 1995

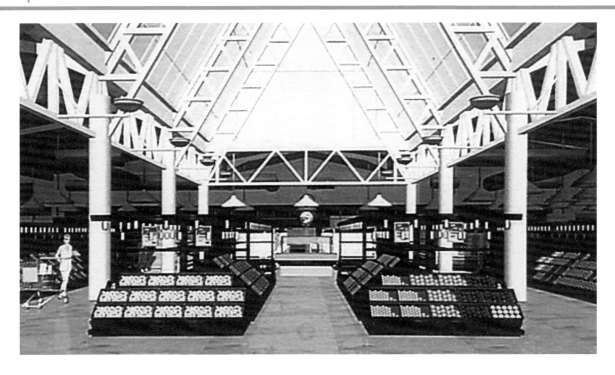

Figure 100. *SONATA Image Retail Asda "Store of the 90's" Image by Mark Edwards of CADAIM © CADAIM*

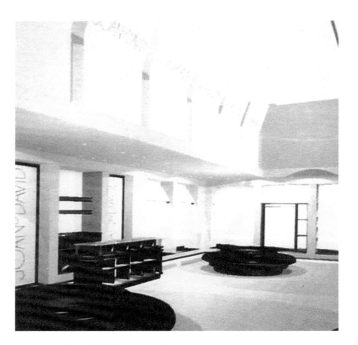

Figure 101. *SONATA Image US Retail Store Layout Joan and David © Giuliano Zampi*

COLES NEW WORLD FIXTURES SCHEDULING [illegible]-may-91
455 WEST LAKES
WEST LAKES SHOPPING CENTRE
SOUTH AUSTRALIA

GROCERY MERCHANDISE

THE FOLLOWING FILE DISPLAYS EACH GROCERY GONDOLA INDIVIDUALLY AND THEN THE RELEVANT MODULE DESCRIPTIONS, CODES AND QUANTITIES ASSOCIATED WITH THAT GONDOLA

1	1430mm WIDE GROCERY ISLAND GONDOLA, 2000mm HIGH 7 MODULES LONG			
1	CEREALS		GIB0292R	6
1	CEREALS	GIA0292R		1
1	JAMS,HONEYS,ETC		GIB0370R	4
1	JAMS,HONEYS,ETC	GIA0370R		1
1	SPREADS		GIB0380R	2
2	800mm WIDE GROCERY WALL GONDOLA, 2000mm HIGH 9 MODULES LONG			
2	CEREALS		GWB0292R	7
2	MUESLI & OATMEAL		GWB0322R	1
2	MUESLI & OATMEAL	GWA0322R		1
3	800mm WIDE GROCERY WALL GONDOLA, 2000mm HIGH 2 MODULES LONG			
3	MUESLI & OATMEAL		GWB0322R	2
3	MUESLI & OATMEAL	GWA0322R		1
4	1330mm WIDE GROCERY ISLAND GONDOLA, 2000mm HIGH 15 MODULES LONG			
4	CANNED FRUIT		GIB0450R	6
4	CANNED FRUIT	GIA0450R		1
4	CANNED JUICE		GIB0340R	2
4	DESSERTS		GIB0460R	5
4	MUESLI BARS		GIB0310R	2
4	MUESLI BARS	GIA0310R		1
4	PVC JUICE		GIB0350R	4
4	TETRA JUICES		GIB0360R	6
4	TOPPINGS & CONES		GIB0470R	3
5	1330mm WIDE GROCERY ISLAND GONDOLA, 2000mm HIGH 19 MODULES LONG			
5	BARLINES		GIB0120B	2
5	BARLINES	GIA0120B		1
5	BEVERAGES		GIB0190R	2
5	BOX & BLOCK CHOCOLATE		GIB0130B	3
5	CANNED & PKT MILK		GIB0200R	2
5	CANNED & PKT MILK	GIA0200R		1
5	COFFEE		GIB0150R	4
5	COFFEE BEANS		GIB0160R	2
5	CONFECTIONERY-MULTI BAGS		GIB0136B	3
5	CONFECTIONERY-PRE BAG		GIB0140B	10
5	LEAF TEA		GIB0180R	2

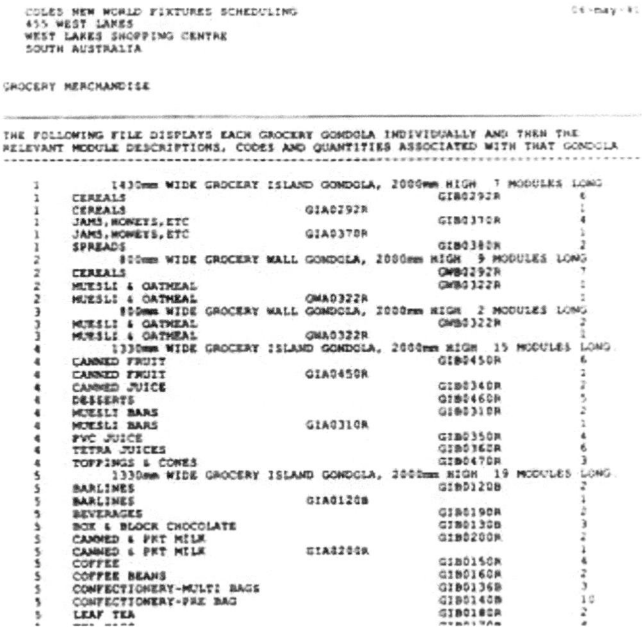

Figure 102. *SONATA Schedule Coles (Australia) Fixtures © Coles Australia 1991*

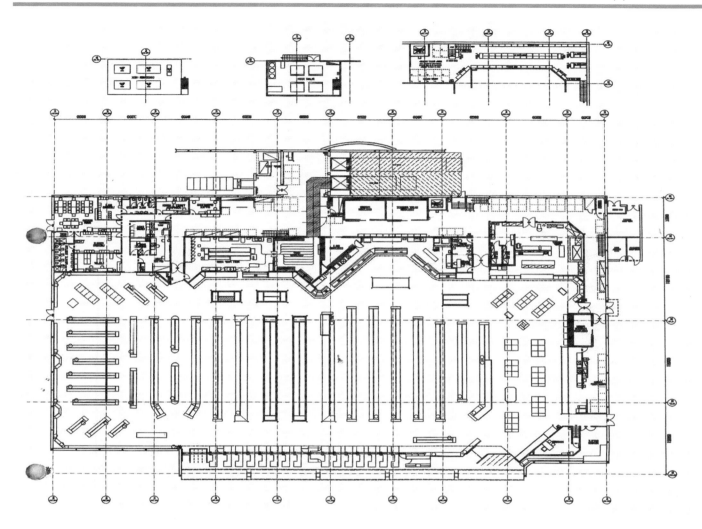

Figure 103. *SONATA drawing Typical Coles (Australia) Supermarket*

the building, were assembled. Bob Wakelam and Lee Danskin produced the store model and images using SONATA.

[Note: The Author continues to work in retail, and Chapter 13, Retail Information Modelling discusses many aspects of modern retailing. The change between his work from the mid-1980s till now (2020) seems vast, but not. The difference is in the viewing technology rather than the principles of Information Modelling.]

CASE STUDY: COLES (AUSTRALIA)

During the 1980s, Coles (Australia) was a rapidly growing supermarket chain. They needed to refurbish existing stores, design new ones and manage the gondolas and goods for sale. In 1988, they installed ten Apollo and one Silicon Graphics

workstation running SONATA at their Melbourne HQ. According to Craig Balding, the CAD manager, these workstations had a program of about six new and twenty refurbished stores a year. They are now one of the largest two supermarket chains in Australia, with more than 100,000 employees.

Coles created a direct link between shelving design and the display of the products, including gondola parametrics to enable suppliers to fit out the supermarket with specific fittings in their correct locations relative to the merchandise. They used a digitizer tablet-based template system for planning, building their own parametric libraries for this purpose. Sixteen drawings per store provided site communication on architectural, lighting, power and services details. Their parametric libraries included diverse

Figure 104. *SONATA image shopping mall Courtesy George Stevenson*

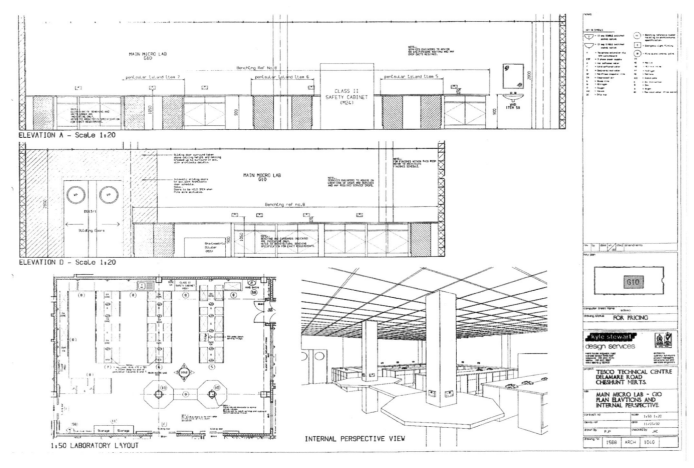

Figure 105. *SONATA Drawing Sheet Tesco Retail Center 1992*

items such as gondolas, equipment and décor layouts, including extensive information for scheduling gondola fixtures and fittings, refrigeration and more general fit-out items. The components themselves wrote information into external databases for further manipulation and break down, so that much of their documentation process was automated.

This early use of BIM helped not just to reduce design and documentation time but increase in the quality level of accuracy and coordinated design detail, making the overall store procurement process more efficient. Apart from the new builds and refurbishment, Coles input many of their existing stores into SONATA for future refurbishments and a Facilities Management database, by digitizing, scanning and modelling the older hand-drawn layouts.

CASE STUDY: ALLIED MAPLES (UK)

Allied Maples, a leading UK carpet and retail store, used SONATA to manage the 180 different outlets they had across the UK. They developed a comprehensive range of fittings and standard components for use in-house using five Apollo workstations and one Silicon Graphics Iris. They expanded the use of SONATA into merchandising and financial planning.

Allied Maples used SONATA to sketch out initial ideas on architectural plans and site surveys. To quote Jacobs (CAD manager) at the time, *"Because SONATA is totally integrated, we can plan in 2D and then bring up the 3D view to see how it will look on site."*

Figure 106. *SONATA Image Men's clothes store Image by Mark Edwards © CADAIM*

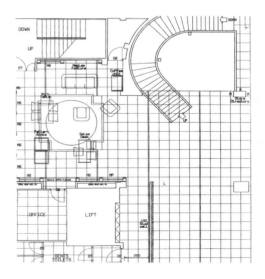

Figure 107. *SONATA drawing Typical Allied Maple Store*

Figure 108. *SONATA Retail Image © CADAIM 1993*

Figure 109. *REfLEX Image Italian Fashion Stores Courtesy and by Massimo Guerini*

Figure 110. *SONATA Image Supermarket 1988 Image by Mark Edwards of CADAIM © CADAIM*

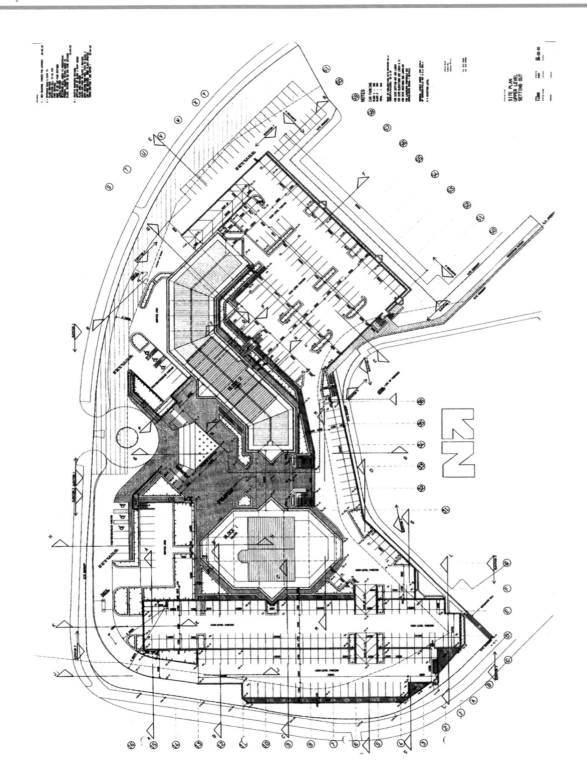

Figure 111. *SONATA drawing Telford Plaza Courtesy Hickton Madeley 1992*

Figure 112. SONATA Retail Image Courtesy Image by Mark Edwards of CADAIM © CADAIM

Using SONATA's parametric capacities, the CAD department generated parametric components based on the style of particular stores and the image they were projecting. To quote Jacobs again *"Maples has an upmarket image expressed by details such as mirrored panels, different colour bandings and column cladding along with ornate stairways and balustrades."* *"With that style held on SONATA, we can apply components directly on the site* *surveys with information on floor space, column heights and wall areas."*[2]

Today, here in the UK, major retailers have gathered together to form the BIM for Retail Group. This collaborative group has the common interest of developing and improving awareness, skill and use of BIM technology.

2   Interview from "Run International CAD journal" Edition 41 published by "The Users Group" 1995

Figure 113. SONATA Image Marks and Spencer refit courtesy Taylor Woodrow & Marks and Spencer

Figure 114. SONATA Image Gas Station design for Spanish Oil and Gas company Repsol © Giuliano Zampi

BIM & Engineering

9

The construction industry is highly competitive and complex, with high customer expectations regarding productivity, budget, coordination and accuracy, forcing professionals to use all means at their disposal to survive. BIM provides an opportunity for engineers, in particular, to meet these challenges and provide competitive products in this challenging environment.

BIM brings a wide range of advantages for Engineers, from control to understanding, from certainty to clarity, illustrated here by the case studies. Clear benefits were obtained that can be seen, even though these were the first attempts at using the virtual information model to solve engineering problems.

The Author trained and practiced as a Civil Engineer, and so both SONATA and REfLEX were written with a distinct engineering perspective. The aim was to allow complex parametric objects to represent engineering structures. This diversity of engineering solutions can be found in these next chapters that include a complete self-designing stadium to the automatic design of cores and staircases of a building, from road design to self-solving duct networks (Services chapter). The Chinese Government was one of the most significant purchasers of SONATA ; they used it to design power plants, power networks and cities. The diversity of these early solutions is enormous and shows the benefits that could be achieved in engineering and services design.

WHAT ARE THE ENGINEERING BENEFITS OF BIM?

BIM brings many benefits to engineering and hence to the building as a whole. Principally these benefits are improved productivity, improved project insight, enhanced collaboration, constant up-to-date information access and enhanced visualization. A more comprehensive list is given at the end of this chapter.

Sharing the virtual model between various stakeholders involves the engineers in their ivory towers with all stakeholders in the projects. This improved collaboration, leads to more realistic and comprehensive proposals. Alternative designs can be tried and costs can be refined, achieved in conjunction with the other stakeholders in full view of the potential operators.

Building a virtual model of projects using parametric elements enables various other benefits. Complexity and detail can be tested, checked and viewed, bringing huge benefits, including reduced rework and construction costs, coordinated documents and information, better cost control, simulation of construction and processes, contributing to better quality projects. Parametric components also allow rule checking, standard and code compliance: even potential self-designing components are possible in an engineering sense.

Building a virtual model also improves the sustainability after examining alternatives, improving safety through an understanding of the hazards, enabling the use of the latest materials and goods with direct external links to suppliers' databases, thus reducing waste. The overall effect is to improve the product, reduce costs, reduce change orders and the resulting disputes, improve profitability and to have buildings completed on schedule. Benefits to engineers also extend to construction management, though this is covered in the next chapter.

These studies show the first engineering projects to use the single unified, coordinated, parametric virtual models. Tunnels, roads, airports, coffer dams, bridges, power stations, prefabricated modular buildings, parametric stadiums, and a water-treatment plant were designed and built using these first BIM systems in the 1980s and 90s.

Summarizing all of the benefits seen in this chapter into a single list might include at least

- Improved collaboration between owners and design firms
- Reduced rework
- Reduced construction cost
- Better cost control and predictability
- Better visualization of information
- Simulation of construct and processes
- Coordination of information and documentation
- More consistent and accurate documentation, so less error
- Determine dependencies
- Improved approval rates
- More exploration of alternative design concepts
- Determine constructability
- Aggregate diverse data into the single model
- Greater flexibility to make design changes
- Improved safety
- Faster, more informed client approval
- Improve designs by building concepts into building components
- Apply standards more easily
- Create more realistic, understandable proposals
- Accommodate late and even ongoing changes to design
- Ensure the latest components and materials through external links
- Reduced project duration
- Reduced change orders
- Prebuild difficult aspects of the structures
- Enhanced marketing image
- Reduced waste
- Monitor as-built structure
- In-situ design calculations
- Increased profit
- Modular design
- Improved order and delivery possibilities
- Whole life asset management
- Rule based checking of component assemblies

CASE STUDY: PARAMETRIC BUILDING CORE, KYLE STEWART

Designers at Kyle Stewart (now BAM), formerly a major UK design and build contractor, created standard parametric core designs for office buildings. These virtual components were self-sizing according to the number of stories, total floor area and space usage, producing coordinated schematic drawings in plan, elevation, section and 3D perspective. It had rule checking, referred to external files for lift sizes, stairwells and other services based upon requirements according to design codes, and included full schedules for costing and construction. The arrangement could be altered rapidly, allowing different building prototypes to be tested. The screenshot shows the lift and specification, the complex

stair detail and an automatically sized bathroom. See Figure 115 opposite.

Complex parametric components within the virtual model not only allow design and calculation checks of complete buildings but also of the smallest details. With such components, coordinated working data, including drawings and schedules, detailed output to specific formats and structures, is guaranteed. For example, data output to CNC fabrication machines, energy analysis, complex surface design, complex data analysis, standards compliance, regulation and rule checking are all easily achieved, to name just a few.

Kyle Stewart had a large team of SONATA users latterly to coordinate and integrate design. Many of their projects required them to work with CAD files from other designers and trade contractors which posed management challenges of working with the most current versions.

Argent, the developer of Brindley Place in Birmingham, gave Kyle Stewart the challenge of trimming the project cost by 10% to make it affordable. Argent had a philosophy of partnering and sharing risk/reward with the whole project team so the cost (not the price) needed to be reduced. Kyle Stewart decided to use Sonata to create a federated model from the designers and trade contractor's files so the team would consider the whole building and not simply their work package.

By 1995, Engineering Technology had been experimenting with the new medium of the Web and applied that to manage the assimilation of and management of files that needed federating – so Sonata became the core of the earliest common data environment, BIW, which is now part of Oracle Aconex. The architecture, structure and services were brought together and co-ordinated into a single model and fly throughs, Virtual Reality and 4D models were generated to improve communications.

Paul Durden, Kyle Stewart's project manager, recalled "*Discussions over buildability were made much easier with SONATA, as we used the complete building model to explore options with the designers. As the design developed, interfaces and the detailing at junctions were considered – raising questions far earlier than is usually the case, and giving time for the best solutions*".

Aspects of work that would normally be considered as part of the fit-out were carried out as part of the main construction, providing considerable cost saving to the overall project. The relationship between Argent and the tenant, BT, was reinforced by the improved communication, reducing misunderstandings and

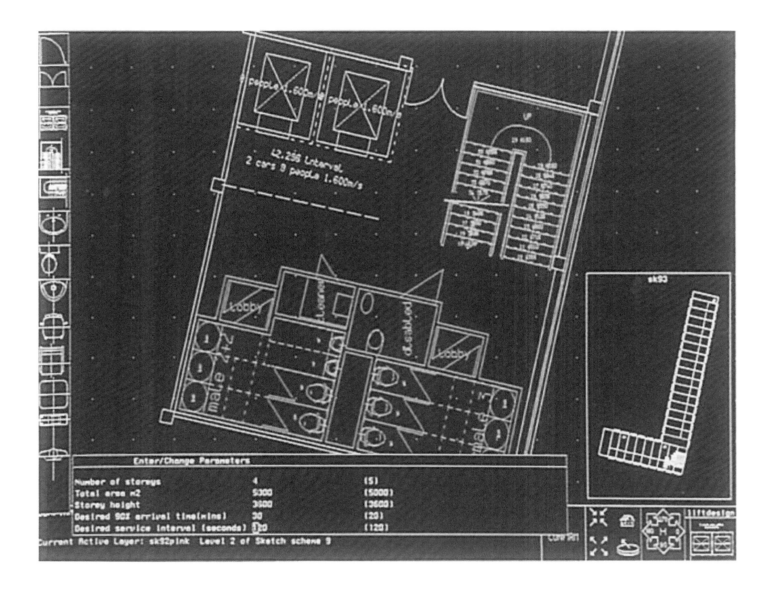

Figure 115. SONATA Screenshot Parametric core design components, Kyle Stewart Courtesy of BAM Construct 1987

last-minute changes. This, in turn, reduced costs and risk on the overall project.

Paul Durden, Kyle Stewart's project manager summarised *"the Sonata model allowed us to pull everyone together, identify with the finished building and focus on delivering a project we all felt we had contributed to."*

CASE STUDY: PARAMETRIC STADIUM
ALFRED MCALPINE AND ENGINEERING TECHNOLOGY

One of the most ambitious parametric elements was a "Self-designing stadium," developed for a highly innovative sports stadium in Huddersfield, UK. The then leading construction company Alfred McAlpine built and partially funded the project using prototype elements that allowed schemes for any chosen site to be generated along with cost estimates[1], developed by Engineering Technology (ET). See Figure 117 opposite.

Parametric elements allowed the stadium *"design to be modified easily and quickly, views and production drawings for any chosen site were generated, along with estimated costs and indications of the quality of accommodation provided. Development so far has covered the general layout of the stadium, allowing the curvature of the building to be explored and indicative quantities to be taken-off."*

"A pitch element compares the intended playing area with current international standards for Soccer, Rugby League and Rugby Union. A variety of structural elements for the spectacular "banana" steel trusses has been created. . . determining the optimum design for maintaining the all-important quality of view from all seats in the stadium, and allowing the client to view the stadium from any seat."[2]

Alfred McAlpine was also heavily involved in cost estimation systems with the University of Salford, together with Engineering Technology Ltd, using REfLEX. Here they used the BIM database to establish structural, electrical and mechanical costs directly from the floor space used alone. They developed expert parametric components that gave an accurate picture of ultimate costs directly from the basic dimensions. The accuracy achieved was 5-10%, similar to what was then achieved through conventional estimating methods. Quoting documents of the time: *"since the information on which the estimation is based determines the*

---

1   "Run International CAD journal," Edition 41 published by "The Users Group", 1995

2   "REfLEX Review", Publication of REfLEX distributors and dealers, June 1995

accuracy rather than the estimating procedure itself. In a sense, the expert system [parametric elements] is only doing what a QS would otherwise do." They found a way to optimize the design using this technique, giving an example of how many variables could be considered, *"such as ceiling height could be altered without substantially affecting total cost."*

They also developed a "module" that generated a detailed BIM "using only limited information." The example given in their Project Modelling in Construction is the design of steel-framed industrial buildings. REfLEX was able to produce *"fully dimensioned and detailed factory buildings simply on the key design parameters. Knowing the portal frame width and height, the program [REfLEX Parametric Components] is able to determine all aspects of the structural elements, cladding and roofing."*

CASE STUDY: TING KAU BRIDGE,
HONG KONG GOVERNMENT

Bridge design and construction bring together several different design areas including roads, lighting, structural engineering, ocean engineering, architecture, programming and simulation work. The advantages of using a virtual coordinated model shared between the different disciplines are clear. The Ting Kau Bridge in Hong Kong was the first substantial bridge built using BIM, was completed in 1998. See Figures 116 and 118.

The 1.18km long cable-stayed bridge is the primary connection between the HK International Airport on Lantau Island and the rest of the city. It was designed by Schlaich Bergemann & Ptrs, structural engineers, and modelled by Giuliano Zampi using SONATA. Separate decks on either side of the three towers contribute to its slender appearance also helping to reduce loading during typhoons. It is one of the longest cable-stayed bridges (as distinct from a suspension bridge) in the world, the

Figure 116. *Photo Ting Kau Bridge*

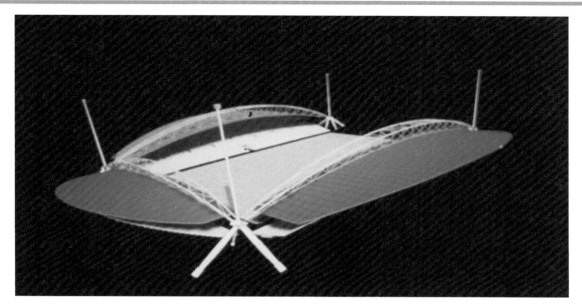

Figure 117. *REfLEX Parametric Stadium Alfred McAlpine and George Stevenson Courtesy George Stevenson 1995*

Figure 118. *SONATA Image Ting Kau Bridge, Schlaich Bergemann & Ptrs Structural Engineers 1993 © Giuliano Zampi*

design and construction cost was HK$1.94 billion, built between 1994 and 1998.

CASE STUDY: HEATHROW EXPRESS, MOTT MACDONALD & TAYLOR WOODROW

The Heathrow Express (HEX) project was a new high-speed railway between Paddington Station and Heathrow Airport. The railway runs along the existing track from Paddington to Hayes, and then along approximately 6.7km of purpose-built track to the Airport.

Unfortunately, some sections of the station tunnels collapsed during construction in October 1994, so a 60m coffer dam had to be constructed to allow safe excavation and reconstruction.

After the collapse, Mott MacDonald, the UK's largest consulting engineering firm, and Taylor Woodrow, then one of the UK's largest contractors, used BIM for work within the stations and the related coffer dam design. Mott MacDonald had written parametric components in-house using REfLEX to create a BIM-based on tunnel design and settlement analysis.

It was considered innovative in 1997, for the first time bringing live data from the construction site directly into the design model in a working environment, superseding desk-based trial and error calculations of the past. The Author assisted Mott in porting the laboratory tested ideas into REfLEX libraries, where a virtual sensor component in the model read live data from an actual sensor in the HEX structure. Any movement in the actual structure was reflected, somewhat amplified, in the virtual model, which was "adjusted" by ensuring that it would regularly regenerate. An alarm system was integrated into the virtual sensor to indicate when changes were being made to the model automatically.

Quoting from a document of the time, other facilities included in the application included:

- Tunnel element including all views for circular-bored, NATM and rectangular tunnels which the Author built within REfLEX for the different tunnel shapes
- Station element with ventilation shafts, platform, lift shafts, escalators and cross passages
- Surface generator element
- Borehole element capable of reading data from borehole logs. The shape of the element in the model changed based on what was happening in the borehole
- Grouting element for monitoring the effects of pumping grout into the earth to stabilize ground movements
- Mapping data import
- Building damages analysis

Instrumentation and monitoring were an integral part of the coffer dam construction, where thousands of instruments were read and the data stored, some of which was accessed by the REfLEX model. The model produced realistic 3D images of certain areas of the project. Still, more importantly, it was also possible to move around the model in real-time and to demonstrate construction sequences, producing time-related prototypes for the project.

Taylor Woodrow also used REfLEX in different parts of the design of the station, including train positioning and phasing of steelwork delivery, the positioning of the new ventilation building in Terminal 4 and other works. The Author was directly involved, providing functionality for NATM tunnel design, the automatic layout of 10km of track and assisting in the design of the BIM components to read sensors from the coffer dam. The virtual model was shared between Mott and TW.

Coffer dam steelwork design was imported into the model from the steelwork contractor's 3D analysis package (STAAD) to assess clash and coordination problems. The model highlighted that a 2-metre high protective wall at the top of the coffer dam would have to be broken through to install the top level of steel, something previously overlooked. Planners were able to incorporate this activity into the program and thus reduce site disruption.

HEX was one of several rail projects developed as part of BAA's Group Rail Strategy. Initially the project was a joint venture between BAA and the British Railways Board. By completion, it

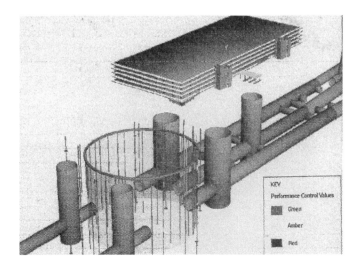

Figure 119. Mott MacDonald Heathrow Express Coffer Dam Pass System Courtesy Mott MacDonald

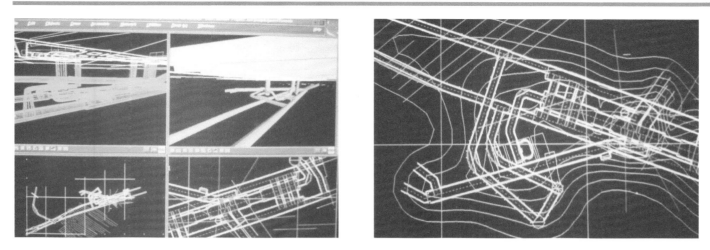

Figure 120. REfLEX Screenshot of Heathrow Express Tunnels © Taylor Woodrow

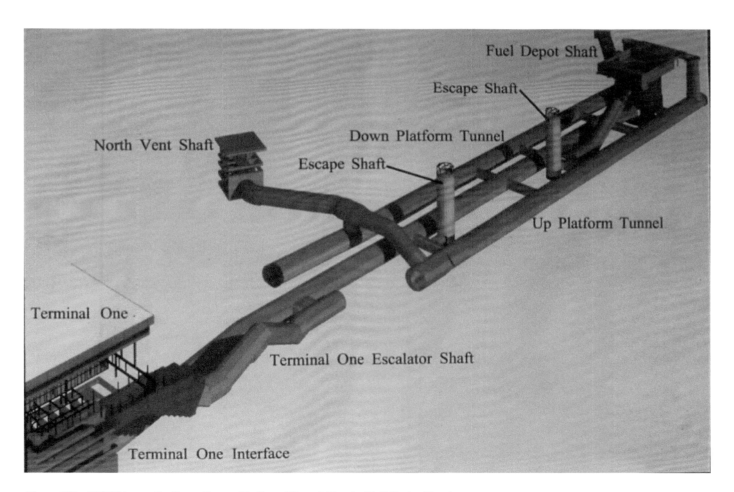

Figure 121. REfLEX image Heathrow Express Shafts and Tunnels Terminal 1 © Taylor Woodrow

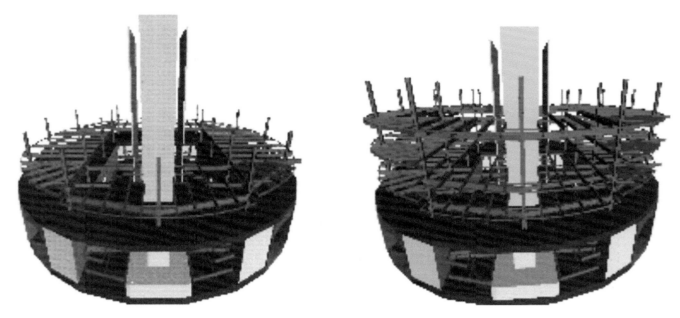

Figure 122. *SONATA Image phased steel work HEX station © Taylor Woodrow*

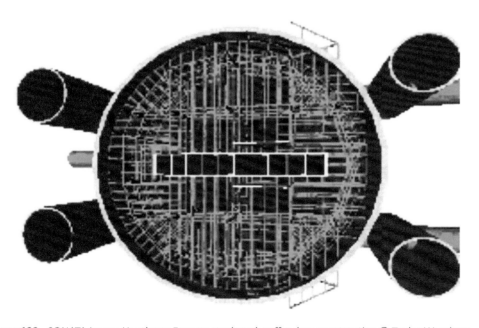

Figure 123. *SONATA Image Heathrow Express steel work coffer dam construction © Taylor Woodrow*

Figure 124. SONATA Image Heathrow Express Tunnels and Steelwork © Taylor Woodrow

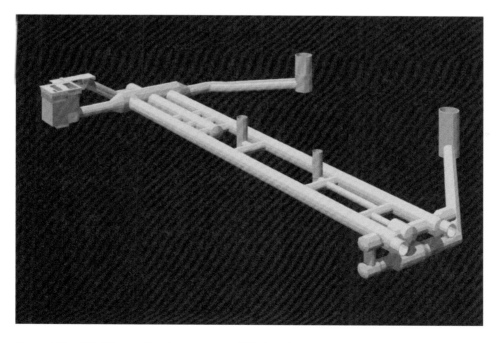

Figure 125. REfLEX Image Heathrow Express © Taylor Woodrow

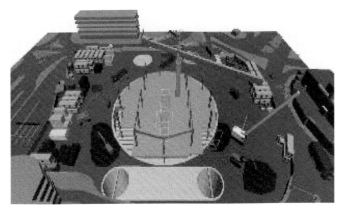

Figure 126. SONATA Image Site Heathrow Express © Taylor Woodrow

was wholly owned by Heathrow Airport Limited (HAL), and is now operated by the Heathrow Express Operating Company under a joint operating agreement with Railtrack.

CASE STUDY: MODULAR STATIONS
TAYLOR WOODROW

Modular, off-site construction, and prefabrication, have been recognized ways of lowering construction costs and improving quality for centuries. From stone masonry to brickwork, carpentry and joinery, it has been going on for years. BIM brings these benefits to today's construction requirements, including technical support, sustainability, safety, constructability, quality, program and enhanced profitability.

The realism especially associated with BIM simplifies simulation and modelling misunderstandings, enabling participants to grasp complex systems' interactions more clearly, particularly in the design workshop environment.

Moving labour off-site, maximizing the use of modular construction and multi-trade prefabrication bring quality, cost and time benefits. BIM can be used to investigate prefabrication opportunities, validate the budgets of modular projects and simulate the installation of these assemblies within the overall project schedule, using lean processes to drive up the reliability, efficiency and profit of the project.

Taylor Woodrow was the first company to use BIM in modular infrastructure to model ground level railway stations in the UK. The study looked at the physical placing and enclosure of standard modules that formed booking offices, stairways, waiting rooms, platform canopies, public toilets, staff accommodation and equipment rooms for three new stations. Existing drawings for the station studies were loaded into the BIM (then called

Building Object Modelling)[3] model, and parametric elements were assembled representing the various construction modules.

The first stage in the process was to model the steel frame using nominal-sized steel members. The frame connections were designed as moment connections to reduce the requirement for cross bracing. These calculations were done within the model.

Curtain wall sections were then fitted incorporating doors, ventilation, vision and opaque panels. Checks were made in the virtual model for sufficient tolerance in the steel frame, curtain wall manufacture and site assembly inaccuracies. Virtual model visual checks were made to check functionality and form.

Canopy roof sections were modelled to ensure they could be lifted in place in 6m lengths. Virtual checks were made to ensure the cantilever brackets on the steel frame via stubs could be weather sealed to each other. Prefabricated 'top hat' sections covered tolerance gaps between the units. The main roof panels were assembled with maintenance walkways pre-fixed.

Models for each station were completed in stages to develop a logical design and construction sequence. A planning element was developed in the form of a 3-dimensional zone object to determine new pod configurations for ticket offices and platforms. Each zone had an assigned function controlling its area and volume; e.g. staff mess room - 15m², waiting room - 10m², etc. After the new pod layouts had been designed using the planning elements, curtain walling objects such as doors, glazed and solid

---

3   "Project Modelling in Construction: Seeing is Believing" Fisher, N, Barlow, R, Garnett, N, Finch, E, Newcombe, R. Thomas Telford 1997, page 2

Figure 127. Amersham Station as built © Taylor Woodrow

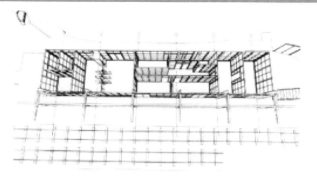

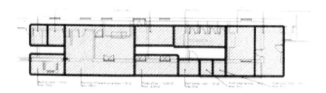

Figure 128. *SONATA Steel Frame of Modular Station © Taylor Woodrow*

Figure 129. *SONATA Plan View of Modular Station © Taylor Woodrow*

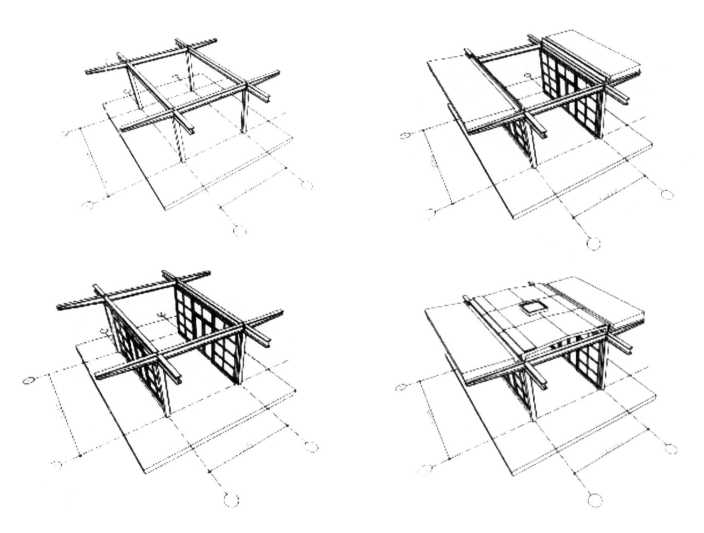

Figure 130. *SONATA Steel Frame through Canopy Roof Assembly of Modular Station © Taylor Woodrow*

Figure 131. *SONATA Exterior Views across Amersham station © Taylor Woodrow*

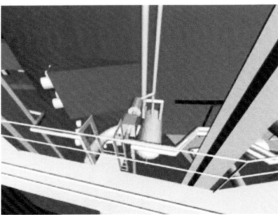

Figure 132. *REfLEX Video, Superheaters within the Reactor Building in an Advanced Gas Cooled Reactor © Taylor Woodrow*

panels, were used to complete the modular layout. The UTS gates with barriers and the attendant's box were located in the main booking hall and the canopy roof modules installed. The study confirmed that virtual construct allowed engineers to foresee many potential issues in the modular design process.

CASE STUDY: POWER PLANTS BNFL

BIM has been used to design industrial structures and Power plants from an early stage. The People's Republic of China (PRC), purchased 72 SONATA licenses through the Ministry of Machinery and Electronic Industries, using it to design and build many industrial buildings and process plants. Unfortunately, examples of their design work could not be recovered. Other PRC departments bought SONATA, including the Ministry of Construction for the Beijing Urban Design Institute and the Ministry of Railways, so many, that by 1994, the PRC was the largest SONATA customer. Thai Energy Administration Authority also used SONATA to design power plants and transmission facilities, including towers and associated buildings. Both the PRC and Thai Governments used SONATA in Urban Planning, unforeseen at the time.

In the UK, SONATA was also used in power generation. BNFL Engineering was an offshoot of BNFL (British Nuclear Fuels Ltd), a nuclear energy and fuel company owned by the British Government. They had accumulated more than 30 years of experience in the design, construction, commissioning and decommissioning of advanced technology process plant for the global nuclear industry. BNFL purchased a number of SONATA licenses and then went on to buy REfLEX, aiming to achieve an integrated approach to the whole design process, including plant

and steel design, 2D drafting and, although unintended, project planning software was integrated. George Stephenson and his company, Engineering Technology, advised BNFL Engineering on this work, for which a brief clip of a power-station model has been recovered.

SONATA was used by BNFL over several years prior to 1995. It initially involved modelling and time-eventing the emerging design of a new large build construct in terms of access, buildability, safety and cost. Ongoing designs were incorporated into the evolving model.

In July 1995, BNFL acquired a Silicon Graphics Indigo 2 XZ configured with both SONATA and REfLEX. The BNFL team were selected and trained in PDMS (Plant Design Management System) allowing design integration across plan and process engineering and civil and structural engineering. Engineering Technologies, who had been assisting BNFL, and the CadCentre (authors of PDMS) looked into building a direct bidirectional link between REfLEX and PDMS.

The huge early take-up by these different organizations, mostly for engineering work and organizations, show that the single, coordinated, parametric, BIM virtual models were ideal for complex applications. Figure 131 shows frames taken from a video of an Advanced Gas Cooled Reactor.

CASE STUDY: MILLENNIUM BRIDGE, BRMBZ

BRMBZ was a group of engineers and architects who pushed the boundaries of design in the 1980s and 90s. They created award-

Figure 133. *SONATA Images Millennium Bridge Thames Design Team BRMBZ courtesy © Giuliano Zampi*

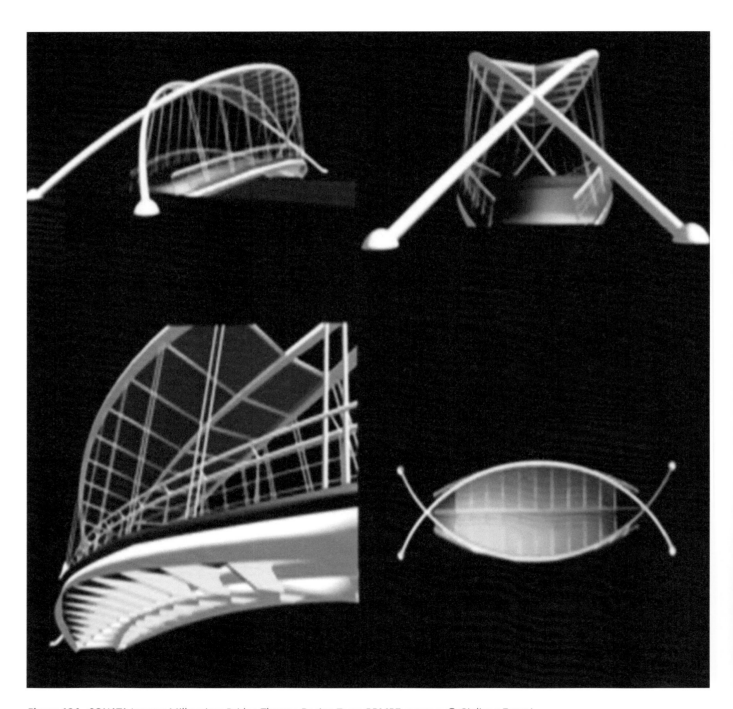

Figure 134. *SONATA Images Millennium Bridge Thames Design Team BRMBZ courtesy © Giuliano Zampi*

Figure 135. SONATA Image, FAC Car park, Melbourne Airport by Jeff Findlay © Peddle Thorp 1988

```
                  Materials Measurement Take Off
                  Work Section D20: Excavations
-----------------------------------------------------------------
PROJECT TITLE.       New Offices.
JOB NUMBER:          J001/91
CAD PROJECT NAME:
TAKE OFF BY:
DATE OF SCHEDULE:
```

Work Section Description	Unit Measure	Quantity	Unit Cost
Excavating to Pits			
2 No. not exceeding 2.00m depth	m3	2	
2 No. not exceeding 4.00m depth	m3	4	
Excavating to Trenches			
not exceeding 1.00m depth Width over 0.3m	m3	45	
not exceeding 4.00m depth Width over 0.3m	m3	18	
Working Space Allowance to Excavs			
pits	m2	27	
trenches	m2	159	
Earthwork Support to Excavs			
max. depth 1m; distance between opposing faces less than 2 m.	m2	36	
max. depth 1m; distance between opposing faces less than 4 m.	m2	63	
max. depth 1m; distance between opposing faces greater than 4 m.	m2	27	
max. depth 2m; distance between opposing faces less than 4 m.	m2	11	

Figure 136. SONATA Ground works calculations (dot-matrix printer) 1991

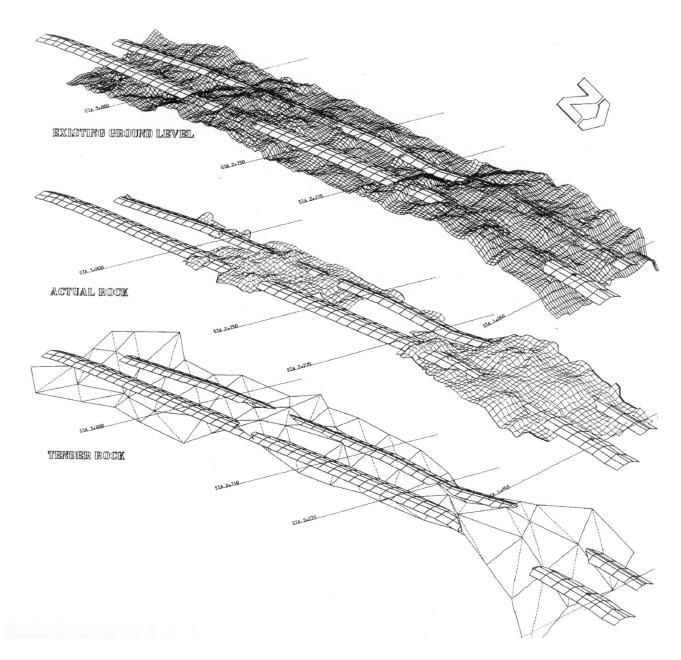

EXISTING GROUND LEVEL

ACTUAL ROCK

TENDER ROCK

Figure 137. SONATA Model, "Ground Modelling" Courtesy & © Frank Shaw and Partners, Chesterfield, UK, 1990

winning designs for several different structures. They created many award-winning designs for a number of different structures. Team members include: Andy Bow, a senior partner at Norman Foster and Partners; Chris McCarthy, Arup's Structural Engineer; Guy Battle, Arup's M+E Engineer; and Alex Ritchie and Giuliano Zampi were the creative designers. The example of their work shown here is of their entry for the millennium bridge competition in 1996. (Figures 133 and 134)

CASE STUDY: ROAD AND GROUND MODELLING

BIM is now used for infrastructure design around the world. It brings efficiency to the production of integrated and complete design documentation, allowing multiple alternatives to be considered. The refinement and the production of coordinated construction documents is achieved more rapidly. Projects are shorter together with more predictable construction timetables. Other benefits include integrated project planning and improvements in highway and road design from visualization, simulation and analysis. Designing for constructability helps reduce these issues and can significantly reduce construction costs.

Integration between the various disciplines also helps reduce issues on site. For instance, integrated lighting, pipe, drainage,

landscape, geo-technical and perhaps even building models, can avoid obvious problems and should enhance the overall design. Automatic input from surveying instruments and ground radar can also be quickly assimilated into the model: see Heathrow Express Case Study above.

BIM facilitates constructability and road safety. Visualization and simulation within the model ensure that road curvature and visual obstructions, such as barriers, berms and foliage, don't affect sight-lines, a crucial part of highway design. The Author's first job was country road design, measuring and laying out roads with steel chains, manually drafting the original layout and later adding the horizontal and vertical curves to drawings. This tedious work ensured that basic facilities for road design were included at an early stage in SONATA's development. Figure 137 shows an automatically generated ground model with added parametric road design.

CASE STUDY: WATER TREATMENT PLANT, WELLINGTON CITY COUNCIL

In 1988, the Wellington City Council (NZ) used SONATA to evaluate the environmental impact of five different proposals for the water treatment plant. Planners and the public could see the impact the plant would have from a number of different locations, including

Figure 138. *Visual Impact study of sewage disposal plant by Murray Pearson*

Figure 139. REfLEX Photo Montage from Robert Grubstad in Sweden

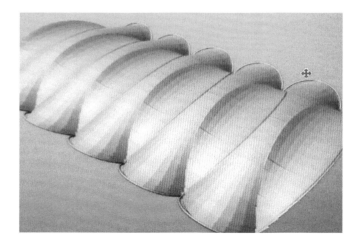

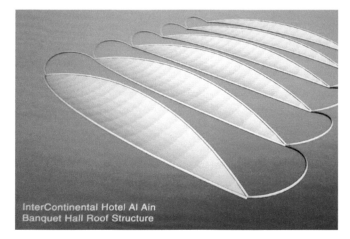

InterContinental Hotel Al Ain
Banquet Hall Roof Structure

Figure 140. ProReflex Parametric roof and tent structures © Peter Dew and Associates 1999/2000

the passenger terminal at Wellington International Airport. The SONATA model covered an area of 15 sq km and included a ground model imported from Beca Carter, Hollings and Ferner. Figure 138 shows the scheme favoured by PWT New Zealand Ltd, for which Murray Pearson made the model and images.

CASE STUDY: INDUSTRIAL PROCESS AND BUILDINGS

The advantages of BIM to process and industrial engineers are numerous, particularly coordination and integration. However, the benefits of full information being available for immediate access by all involved are considerable, including, for example, facility engineers who may need a process valve schedule and drawings that match valve tags installed in the field. The facility engineer can find the valves in a BIM, where they are modelled, installed and tagged according to the facility management standards.

As point of possible interest, having designed various pipe networks by hand while studying and as an engineer, the Author wanted to emulate PDMS, (Plant Design System) and worked hard to make SONATA strong in this area. There was a time when this was the intended market.

Another example of an industrial BIM application would be a central equipment plant upgrade, where piping and ductwork may need to be reconfigured within a defined renovation area and connected into building infrastructure. This is easily done in a virtual model. The possibility of feedback into an engineering model from different systems within the plant or factory may need to be considered, achieved in monitoring the coffer dam in the HEX project. All of the usual BIM advantages are already available, including clash detection, etc., for which a good example of industrial plant modelling was the Svenska Stat Pol refinery: see Figure 139.

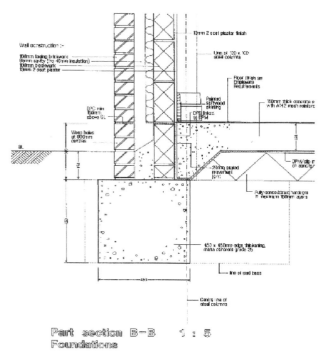

Figure 141. *SONATA detail 1992*

Figure 142. *SONATA Steel Frame Design Images by Mark Edwards of CADAIM © CADAIM*

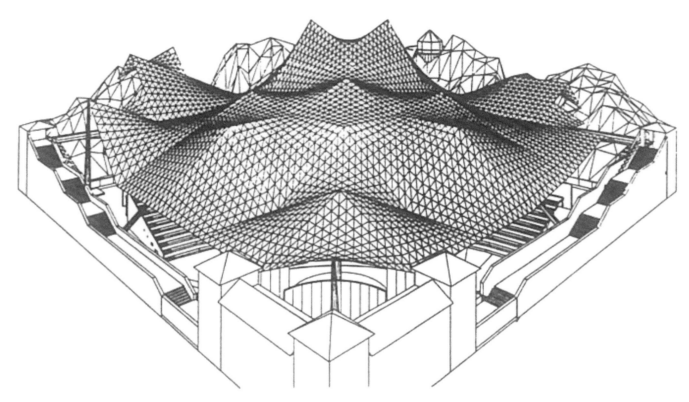

Figure 143. *SONATA Parametrics used for hanging roof at the London Zoo 1991 © Scott Brownrigg*

Figure 144. *SONATA Image, FAC Car park, Melbourne Airport by Jeff Findlay © Peddle Thorp 1988*

BIM & Construction Management

10

Virtual models are an ideal framework on to which to add time-related information, bringing a temporal sense to building design and construction. This information might include lead-times, time to become operational, the construction sequence and time to install. The case studies and examples in this chapter illustrate some of the applications of "temporal" BIM. These can be viewed as an evolving building under construction. Modelling in this way has led to better, inter-discipline coordination, optimal resource planning and dynamic viewing of the model over time.

Using a BIM model to assist in construction management of the building has many benefits:

- Accurately modelled costs based on all component information
- Pre-build to ensure no clash issues
- Delivery to site just in time
- Sequential construction issues predetermined
- Avoid delays in the construction process
- Determine on-site issues
- Explicit information to all stakeholders
- Improved sharing of information
- Improved team engagement
- Accurate and conclusive project closeout
- Improved logistics
- Accurate estimating
- Accurate and fast scheduling
- Improved pursuit and marketing
- Improved information flow around project teams
- Better sharing of results
- Greater management buy-in
- Improved display of capabilities
- Improved awareness of on-site safety
- Clash free construction-scheduling

Comprehensive parametric models of virtual buildings mean costs can be dynamically estimated based on component parameters such as quantity, volume, colour, or supplier. Model-based costing is crucial, and access to files outside the BIM opens up the possibility of powerful cost-based systems. Before BIM, budget costs were either prepared as a completely separate exercise associated with sketch drawings and specifications or based on a final set of documents produced too late to drive the design. Multiple copies of data led to different costing outcomes depending upon which files were used while things were changing, especially during the design process. The rough order of magnitude estimates produced under older systems are now avoided with dynamic breakdowns derived directly from the BIM properties. Client's budgets are far more likely to be honoured.[1]

Planning the delivery of goods to site and storage of materials, temporary works offices and general site planning can be achieved at an early stage. Gantt charts linked to the dynamic operation of items within the model lead to precise planning. Visualisation of partially constructed buildings gives an understanding impossible before, with an accurate temporal view of different aspects of the building.

Administration of the construction process can also be improved; greater document control, clearer information clarification, more realistic project team training, greater safety and improved information processing times.

BIM visualisation with time is perhaps the most valued aspect of virtual models, allowing everyone to consider alternatives before being forced to by circumstance. Change order reduction by 40%

---

1   "BIM Handbook: A Guide to Building Information Modelling for Owners, Managers, Designers, Engineers, and Contractors", Eastman, C., Teicholz, P., Sacks, R., & Liston, K., Hoboken, New Jersey, John Wiley & Sons, 2008

Figure 145. Taylor Woodrow Cover of 1995 brochure illustrating full 5D Modelling. © Taylor Woodrow 1995

from clash detection alone, and up to 90% reduction in variations and rebuilding was achieved.

The very first projects run by Taylor Woodrow were aimed at producing zero change orders once the model was signed off. These first clients understood exactly what was possible[2] from the system.

_____

2   "Project Modelling in construction… seeing is believing", Fisher, N., et al, Thomas Telford Books, 1997

This chapter gives case-studies of the first uses of BIM in construction management, including costs and construction sequencing dates and times. Costs, construction dates, delivery dates and other associated information were assigned to components in the building model. Various tools for viewing and adjusting were provided to watch the construction of the building dynamically and direct links into a Gantt chart to allow live updates to the chart (see Figure 167).

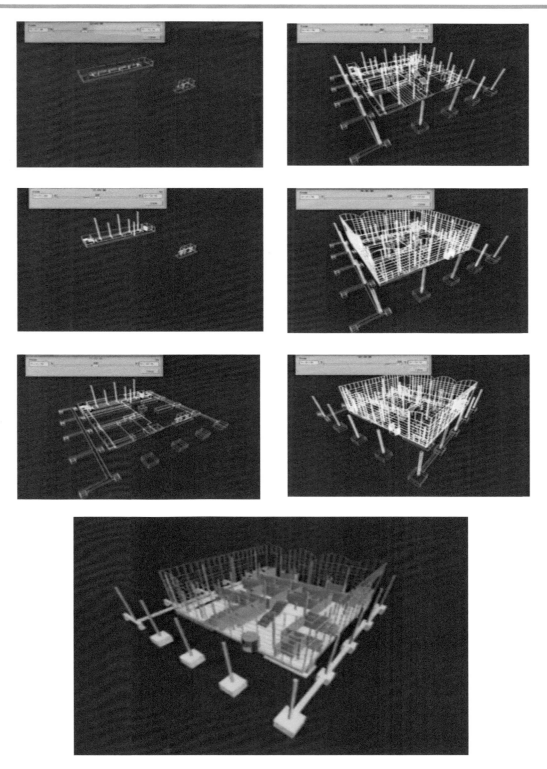

Figure 146. *REfLEX Screenshot Dynamic slider bar with associated Construction Start and End dates*

Figure 146 shows a series of images generated by moving the slider bar, where moving the bar backwards and forwards enabled the user to see the construction phases, with flashing items indicated under construction.

The Author worked full time for a period as a consulting engineer for the construction management team of Taylor Woodrow during the development of REfLEX. This gave a great opportunity to see the possibilities and issues first hand, and to tailor the BIM systems accordingly: much of the development in this chapter stemmed from this work. (This was also done to pay REfLEX salaries!)

"Project Modelling in Construction... Seeing is Believing"[3] was published in 1997, describing SONATA and REfLEX and how they were used on various projects. Fisher calls the model Building Object Models (BOMs) as this was prior to the advent of the term Building Information Models (BIMs). Object modelling is still used to reference BIM in some papers. In the 1980-90s, the Author used the term Building Object Modelling to refer to the single project model, known as BIM since 2003. BOM has been renamed BIM. The hardware supplier Silicon Graphics marketed the O2 (a super machine of the time on which REfLEX was sold) workstation with the term BOM.

CASE STUDY: SMITHFIELD MARKET, TAYLOR WOODROW

Taylor Construction Management was contracted to coordinate and plan the building of a complex new structure within the historic Victorian shell of the Smithfield Market, London. This was the first use of BIM to coordinate information from a range of designers and contractors and to manage the construction process itself.

Engineering Technology, with George Stevenson, used SONATA to coordinate and plan the works, where Taylor Woodrow viewed it for temporal and construction management. Rosser and Russel used SONATA as well, some of the design work itself was done using other systems, imported into the BIM, including drawings and data from HLM (GDS Architecture), Atkins (AutoCAD M&E), Rowans (StruCad-Structural Steel).

Taylor Woodrow Management had also invested in SONATA Modelling technology in 1992 for the Smithfield East Market

---

3   "Project Modelling in construction... seeing is believing", Fisher, N., et al, Thomas Telford Books, 1997

Management Contract. Its use was limited to the coordination of listed ironwork and new steelwork, together with visualisation. See Figures 147 through 150. By 1997, working on the West Market using the same technology, they had brought certainty to design development and construction management of the project. BOM (BIM) had been used as a coordination tool to highlight design clashes and ensure that the Works Contractors received coordinated information. This coordination was key to the COM and subsequent BIM success.

In particular, the British Rail Tunnel that runs underneath the Market had to be strengthened before the additional steelwork to support the two new floors within the structure could be constructed. Modelling was used in the negotiations with British Rail to ensure that the possession periods for the tunnel adequately reflected the work that needed to be carried out. These images had also been used for the HSE to develop the temporary works. Tendering contractors had been issued with a pictorial method statement so that their price reflected the work that needed to be carried out. The model was also used to assist in the phasing and programming of the works.

The key benefits as listed by the project management team of the time were Contractor briefing, Stakeholder interfaces, the certainty of the project outcome, and the what-if studies.

Ray Barnes, ET's software director, was the first to use the term 4D associated with project modelling. According to George Stevenson, *"Ray, whose degree was in Astronomy and loved science fiction, said time is the 4th dimension so we should call it 4D."*

Figure 147. *Smithfield Market original steel work 1868 © Taylor Woodrow*

Figure 148. *Smithfield Market refurbishment in progress © Taylor Woodrow*

Figure 149. *Stages 2 and 3 of the support to the existing columns © Taylor Woodrow*

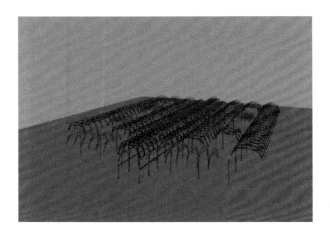 

Figure 150. *SONATA Screenshots Smithfield Market Image Bob Wakelam © Taylor Woodrow*

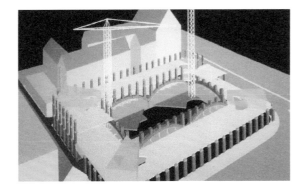
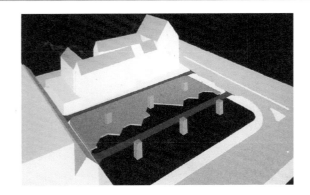

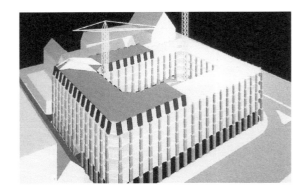
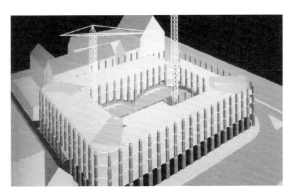

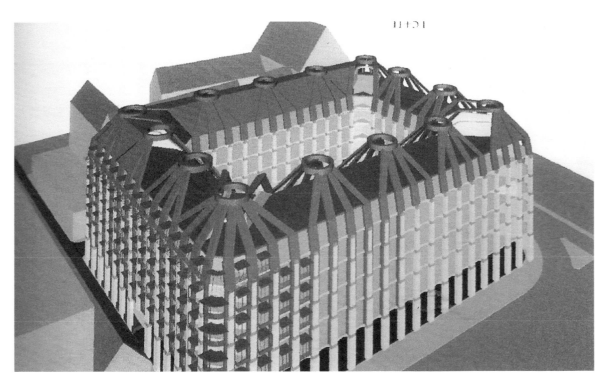

Figure 151. *Sequential views of Construction Phasing of Portcullis House © Giuliano Zampi*

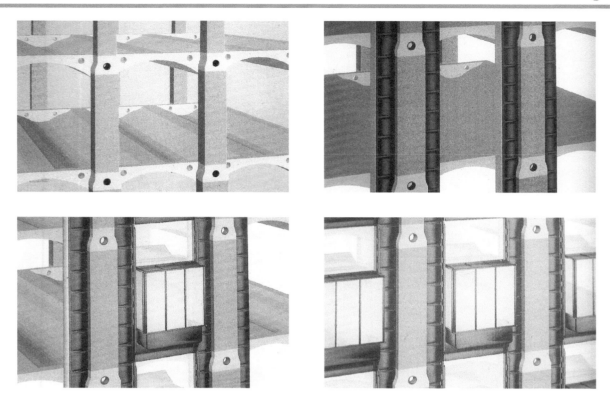

Figure 152. *SONATA Image of Window Detail Construction Phases Portcullis House © Giuliano Zampi*

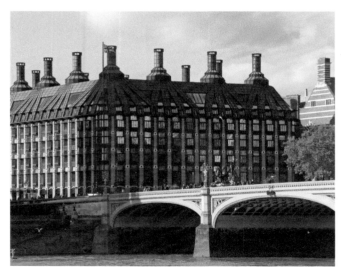

Figure 153. *Portcullis House as-built, Westminster by Tony Hisgett License: CC BY-SA 3.0*

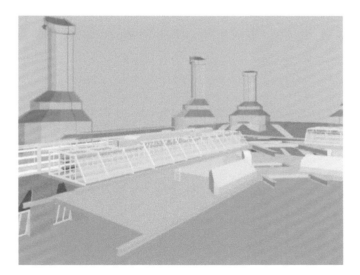

Figure 154. *SONATA Image M&E Coordination of Portcullis House Courtesy George Stevenson*

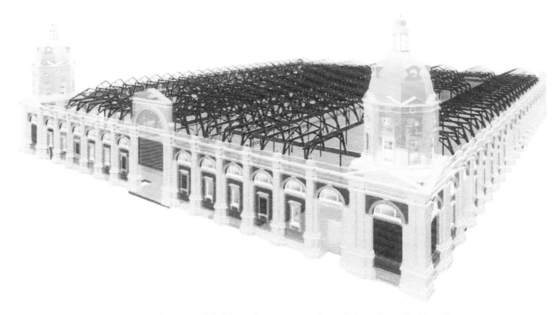

Figure 155. SONATA Screenshots Smithfield Market Image Bob Wakelam © Taylor Woodrow

CASE STUDY: PORTCULLIS HOUSE WESTMINSTER MICHAEL HOPKINS AND PARTNERS

Michael Hopkins and Partners designed Portcullis House, a prestigious office development immediately beside the Houses of Parliament, Westminster, central London. Rated BREEAM Excellent, offices for 210 MPs and a suite of Select Committee Rooms were completed in 2000 at a cost of £165m. Westminster Underground station occupied the same plot area, but well below ground level, designed by others but coordinated at the same time.

SONATA was used to model the final architectural design and program the work, including detailed construction phasing of the building. The impact of any changes to the plan could be determined immediately, saving precious time on site. Critical path analysis was developed for just in time deliveries using the phased model, reducing road obstructions. Electrical layouts were coordinated with lighting, and external planning view lines were all constructed as part of the model. Accommodation is housed around a central courtyard in a six-story rectangular block, covered by a frame-less glass skin, supported by an oak and stainless steel diagrid.

Taylor Woodrow commissioned a model for their bid for Portcullis House but when Laing Construction won it, George Stevenson sold the idea of Project Modelling to MD Brian Zelly, who then used SONATA to coordinate different aspects of the project.

CASE STUDY: EUROPIER TERMINAL ONE HEATHROW, ENGINEERING TECHNOLOGY

This case study outlines the EU-funded BRICC (Broadband Integrated Communications in Construction) study in the early 90s on the BAA Europier extensions in Terminal One Heathrow.

Heathrow was, at the time of this work, the world's busiest international airport (now Atlanta). Currently it accommodates over 100 million passengers each year. Its owner, BAA, invests substantial amounts in the development of the facilities and maintenance, much of which involves construction in areas that are constrained by existing structures, regulations, passengers and aircraft.

Heathrow's runway 5/23 was coming to the end of its operational life, with changes in regulations regarding the proximity of runways and obstructions, and issues of increased parking bay requirements for modern aircraft: solutions involving runway and building modifications had to be devised.

BAA investments *"had to be innovative in terms of their planning and construction. The development of Europier and its associated infrastructure is an excellent example of how the airport's planning and construction teams were able to work together to significantly improve levels of service for Terminals 1 and 2."*

Changing status of runways 5/23 from instrumental to visual, allowed moving the centreline 15m away from the terminal and discontinuing runway 5. Runway 23 was reduced in length owing to the change in status, allowing Boeing 747 stands to be built at Terminal 4.

As part of this redevelopment, the jetty was to be extended to deal with larger aircraft, and with changes in the regulations, arriving and departing passengers had to be segregated, requiring a new building.

Europier is 284m long x 16.2m wide, allowing ten aircraft stands and departures level seating for 1140 passengers. The building has a vaulted roof supported on pairs of steel trees; the rooftop plant was re-sited at apron level to maintain sight-lines from the control tower, and passenger segregation was achieved with arrivals through galleried walkways.

The challenge was to construct this building amid the busy airport while meeting the project procedures, functionality and cost. It was done in part by using the Europier 3D Modelling Project BIM in a fully coordinated, project management sense.

Richard Rogers Partnership and Bovis Construction worked with Engineering Technology, using SONATA as the project-modelling tool for the Europier development. The application of the modelling approach was carried out within the ambit of an EU-funded R&D project, BRICC (Broadband Integrated Communications in Construction). The BRICC consortium was formed by a group of leading European construction and telecommunications companies to improve inter-disciplinary communication and team integration on large construction projects.

SONATA was used on the Europier project to coordinate drawings from different designers and contractors, to check for clashes between the different drawing sets, to view the structure in coordinated 2D and 3D, to simulate the phasing of the construction and to allow simultaneous, globally dispersed engineers to access and manipulate the design.

SONATA was used by only two of the firms involved in the project. The others, using conventional CAD systems, were required to provide their data in the form of tightly specified 2D DXF drawing files. All of the CAD files and drawings used on the project were uploaded into a central project information Hub, where they were reviewed for accuracy and conformance with the agreed standards, and when approved were made available

for use to approved recipients. All of these processes, including management of the central SONATA model, were controlled using Bovis' proprietary Hummingbird Information Management System, operating on a central hub and delivered over an ISDN-based private extranet. After some initial teething problems this system worked very well, enabling technicians to take the 2D files and build them into the model as 2D views of the respective 3D components.

Data stored in the Hub and used in the modelling work included both consultants' design information and contractors' fabrication and shop drawings. Cladding, steelwork, blockwork and building services were all included. Management of designers' and contractors' data within the model was an important aspect of the project, highlighting several unexpectedly complex issues relating to the phased integration of consultants' with contractors' information on such complex, fast-moving projects.

Europier was one of the first live projects to demonstrate the benefit of creating a 3D model of a complex building as part of the design/construction process. It showed that BIM systems such as SONATA could be highly effective in the primary design process, particularly as a means of visualising complex spaces and the features that make them up. An almost trivial example on Europier was the ease with which design and layout of signs were seen and passenger way-finding simulated.

SONATA was used very successfully for many operational purposes, including clash checking, construction simulation, and visualising construction activities in logistics and safety exercises. These sorts of activities, impossible using conventional tools, were all made possible through the use of the clash detection and 3D imaging capabilities of BIM. SONATA also provided a globally available model of the project, enabling various remote participants to examine and modify the design, subject obviously to protocols and control.

The BRICC / SONATA work was initiated somewhat later on the Europier project than would generally be desirable after most of the key construction contracts had been awarded, and much of the technology – ISDN networking for example – was unproven and difficult to procure. However, with very active support from the BAA project team, the consulting firms and the key contractors, a highly constructive, collaborative atmosphere developed around the central Hub and BIM. At least partly as a result of this, both the physical Europier project and the BRICC / SONATA element were subsequently assessed as being remarkable

Figure 156. *SONATA Images Europier Project Model Courtesy George Stevenson*

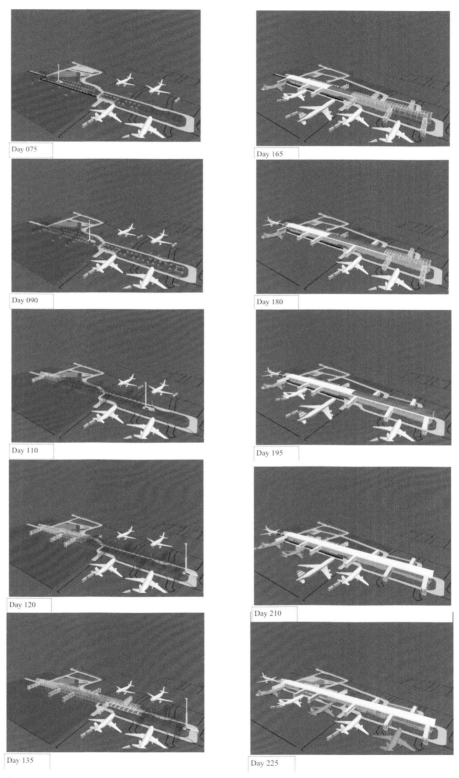

Figure 157. SONATA 5D Europier Project Model Courtesy George Stevenson

153

British Construction in Industry Award
1996

Figure 158. British Construction Award for Europier Design with BRICC

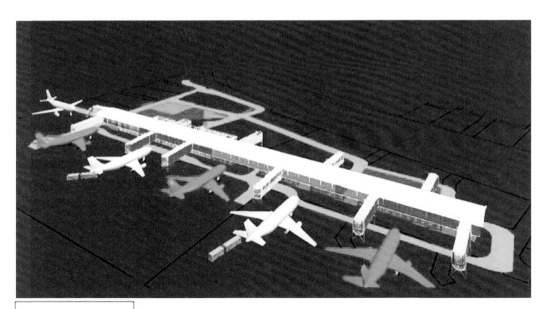

Day 225

Figure 159. SONATA 5D Europier Project Model Courtesy George Stevenson

Figure 160. *Europier as-built 1996. Runway 23 parallel at bottom © ICE Publishing 1999*

Figure 161. *NE corner of Europier pier showing segregated arrivals © ICE Publishing 1999*

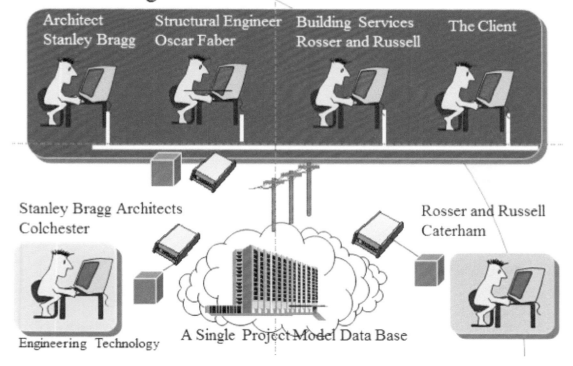

Figure 162. *Slides from 1991 SONATA Shared Project Model Courtesy George Stevenson*

successes, each in its way. Quoting from the Europier Final Technology Report:

"The unique Multiple Project Access functionality within SONATA allows a multiple of users to access the model simultaneously, to quickly solve problems..."

"Another commonly held misconception is that the disc capacity for a modelling system needs to be massive. This was not the case; a SONATA project with all the necessary sheets and drawings created uses far less disc space than the same project on a file-based system."

Quoting from a paper at the SONATA and REfLEX Users Group Conference in 1995: *"BAA's Europier with Richard Rogers and Bovis has provided the real evidence to support the theory that using SONATA to test the design and construction data on a project is self-financing..."*

In 1996 Europier won the British Construction Industry Award – Major Buildings – for excellence in its overall design and planning. Europier has been the lowest cost of any BAA pier of comparable quality. Construction began July 1994 and the first flight was 4th December 1995.[4]

---

4 "Project Modelling in construction... seeing is believing", Fisher, N., et al, Thomas Telford Books, 1997

Before the North Terminal Domestic Facility was given the go-ahead, everyone involved in the scheme, from the designers and contractors to the promoter and end-users, was asked to sign a 'Vision 2000' charter which committed them to make no changes after an approved date. This was only possible because they could see just how the building would appear, work and be built on a walk-through 3D computer model.

The model identified some problems with the original design, which would not have been evident from 2D drawings, leading to changes to the final design. The model was critical for the success of the Design Agreement or 'D-Day' no changes meeting, attended by 45 stakeholders, designers and builders at Gatwick to make the commitment. There were also people from British Airways, local fire services and baggage handling, as well as comments from customs, the police and immigration. Leading up to D-day, there were a series of workshops with all the stakeholders involved. They knew what they were going to be signing. End users were walked through the model and went through a final signing off ceremony.

BAA had what it wanted: signed and witnessed agreements that the project would proceed without changes, giving it the predictability that is needed to achieve its aim of cutting project costs by 30% and ultimately 50%. The key project modelling benefits of the time were, according to Taylor Woodrow:

Figure 163. *REfLEX Gatwick © Taylor Woodrow 1996*

- Feasibility studies
- Material and finishes approval
- Stakeholder decision making
- 'What-if' studies completed

The same model was used to educate the public and in the design of passenger flow.

On completion, BAA attributed the project's success to using "object modelling," now known as BIM, and especially to the use of a consistent project model enabling proper project control.

One of the reasons the Author decided to sell REfLEX rather than continuing to market it, was because even after the great success of this project, the design of Terminal 5 at Heathrow was awarded to a consortium that used a now extinct 2D system developed by INTERGRAPH.

CASE STUDY: ROYAL ALBERT HALL, TAYLOR WOODROW

The Royal Albert Hall has its origins in the Great Exhibition of 1851, which was held nearby in Hyde Park. The Exhibition was the brainchild of Prince Albert the consort of Queen Victoria, Sir Henry Cole, a senior civil servant, and other members of the Royal Society. The Great Exhibition was a resounding success, spurring public interest around the world in British and Commonwealth cultural and engineering achievements, and remarkably, it made a profit.

The money was used to help purchase the area of land opposite the Exhibition site, between Kensington and Cromwell Roads. The aim was to create a permanent equivalent of the Exhibition; a grand area in which to house and display the best of British arts and sciences in perpetuity. The concentration of learned bodies currently located in this area is proof of the success of that ambition. Institutions in the neighbourhood include Imperial College of Science, Royal College of Art, Victoria and Albert Museum (where the documents on which this book are based are now stored), Science Museum, Royal College of Music, Natural History Museum and Royal Geographical Society. The Royal Albert Hall, initially called the Hall of Arts and Sciences, was established in this site.

Queen Victoria laid the foundation stone in May 1867, commemorating the death of her recently deceased husband. Measuring 219 by 185 feet, the Albert Hall is a brick building, comprising about 6 million Fareham Red bricks with about 80,000 buff-cultured terracotta blocks providing a decorative contrast. The roof a double-skinned dome of iron and glass. It is a Grade 1 listed

building, having been used continuously since its opening in 1871, the home of many major cultural events to this day.

Intensive usage and the general ravages of time had led to progressive deterioration of the fabric of the building. In 1987, pieces of terracotta, loosened by frost, fell from the underside of the smoking gallery, the last straw prompting the governing Council of the Hall to initiate a major renovation of the structure and services. In 1995 Taylor Woodrow was appointed to carry out the refurbishment, a £70M programme, performed under a Construction Management form of contract.

"The brief called for visual project-modelling techniques to demonstrate the extent of each trade package and to determine the effects of their work on the operations of the hall and adjoining buildings, users and occupants." Taylor Woodrow proposed the REfLEX 5D project modelling system - three dimensional images allied to cost analysis and time sequencing, advising *"We can simulate all aspects of a project before we even set foot on site."*

The first application of REfLEX on the project was with work on the extension of the balcony corridor and seating. Survey information was used to create a detailed model of the balcony area, allowing designs to be produced and tested virtually, off-site, which ensured that subsequent construction issues were minimised.

"Object-oriented modelling [BIM] was used very successfully on the balcony works to co-ordinate design and spatial information and to create visualisations which were used in workshops and 'dress rehearsals' with the workforce, where the proposed works had been time-evented and planned in advance"

Figure 146 shows the project in various stages of development, where components had construction start and finish time associated with them: REfLEX had a slider bar to move backwards and forwards, to watch the BIM grow or shrink, in full colour. The Author wrote a further REfLEX time control component to look at the database and display a continually updated Gantt chart.

The most substantial part of the project was the development of the south steps of the Hall, which involved the removal and renovation of the Albert Monument, removal of existing underground car parks and construction of new basement levels, all coordinated, tested and simulated using BIM.

The model was also used very effectively for demolition procedures, to design the layout of site operations, to plan construction sequencing, to create visualisations of traffic moving into the service bays, including turning circles and so on, and in setting out the security cameras with accurately planned fields of vision. A video clip from it is still in use today to publicise events at the Royal Albert Hall, the BBC Proms especially, with sequences regularly televised worldwide. This clip is available with the UnderstandingBIM app available in your app store.

As with other BIM projects at the time, design and shop drawings were obtained from various originators in the form of files (.dxf). These were then tested for accuracy and coordination by loading them into an intermediate model which was subsequently checked, discipline by discipline, against the assembled model.

The first steps in the project were to aid in planning the site setup. Hoarding schemes created in the model were used for a variety of purposes: to verify boundaries, to check the viability of potential access and storage locations, and to ensure that the hoardings were arranged in such a manner as to minimise obstruction of the views of Albert Court and Imperial College from pavement level. The ability to present these and similar issues to local residents in the form of an intelligent 3D model, as opposed to the traditional medium of drawings and sketches, was essential. The model provided a vivid, realistic representation of the logistics and building works that were particularly effective in alleviating the residents' fears about the impacts on their lives of the planned construction.

Walk-throughs, animations and colour stills, as well as conventional 2D drawings and sketches, were all produced from the model. Virtual Reality exercises were generated for the client team by Taylor Woodrow. These helped the client appreciate the scope and detail of the proposed work by enabling them to move in a dynamic, un-programmed way through the model, now known as Virtual Reality, derived directly from the content of the building information model. Figure 165 is a view of the VR model used at the time.

Many unorthodox forms of media were included in tender packages issued to contractors bidding for construction packages. A specific example of this was with the Albert Court model, developed primarily to investigate alternative methods of providing a temporarily suspended walkway for access to the building. Once the preferred method was decided upon, details were extracted from the model and were sent to a specialist fabricator/contractor for detailed design, manufacture and installation.

Taylor Woodrow modelled the layout of the foundations from the original construction drawings, used for spatial coordination of a number of new and old elements of the building fabric before demolition and site construction began. The model threw up several issues which were detected and resolved early and economically, avoiding the potentially complex and costly challenges that might have arisen if they had been discovered later, in the course of the construction works.

In the south steps area of the site, where a large greenhouse structure was known to have been located in the past, the concern was that unknown, complex foundations and other underground obstacles associated with the greenhouse might be found, which might impede the planned works. A detailed geophysical survey, using ground-penetrating radar, was carried out by the Museum of London Archaeology service. This broke down the general area into discrete strips in which the extents and depths of different materials were recorded as they were surveyed, and information entered into REfLEX to model the conditions underground. This application was highly effective in identifying and mapping, in three dimensions, the exact locations of steel, concrete and other underground obstacles. This allowed subsequent construction to be accurately planned, carried out efficiently and economically.

Another interesting application of REfLEX on the Albert Hall project was sight-line checking. Every one of the hundreds of curved seat sections arrayed around the Hall is built on a slightly different curve. Each of these sections was individually modelled in the system, then tracked when it was removed, re-furbished and re-upholstered, and subsequently re-inserted in its correct position in the building. Checking sight-lines from different seats was part of this process. See Figure 168.

Virtual Reality modelling language file format videos (.WRL format) were also generated by Taylor Woodrow for the three lower levels to enable the client to move through the model. These files, shown in a modern browser, in the Figure 165, were the first use of VR within any building information model (BIM).

In conjunction with Tony Gee, Consulting Civil, Structural and Geotechnical Engineers, the Royal Albert Hall South Steps work won the Fleming Award for Excellence in Geo-technical Design.

Figure 164. *Royal Albert Hall, photo by DAVID ILIFF. License: CC BY-SA 3.0*

Figure 165. *SONATA VMRL file (Virtual Reality Modelling Standard) RAH-South © Taylor Woodrow 1996*

Figure 166. *REfLEX Image Royal Albert Hall © Taylor Woodrow*

1. 01-02-96 Corridor strip-out.

3. 24-06-96 New seating zone.

2. 01-06-96 Seating strip-out & new seating zone.

4. 21-12-96 Balcony works complete.

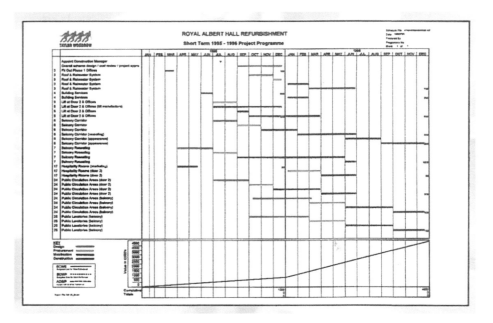

Figure 167. *REfLEX Images Construction Stages and linked Gantt Chart © Taylor Woodrow 1995*

Figure 168. *REfLEX Images Taylor Woodrow Royal Albert Hall sight-lines to stage © Taylor Woodrow 1996*

Figure 169. *Views of Construction Site at South Steps from Prince Consort Road © Taylor Woodrow*

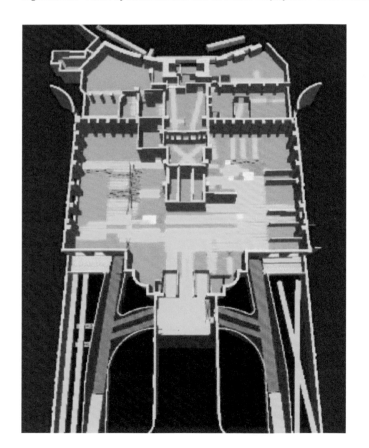

Figure 170. *SONATA Image Ground Penetrating Radar Survey RAH South Steps © Taylor Woodrow*

Figure 171. *SONATA Movie Sequence Entrance to RAH underground car park © Taylor Woodrow*

Mechanical & Electrical Services

11

The challenge in Mechanical, Electrical and Plumbing (MEP) services engineering is to create ideal environments for the users of buildings and for processes carried out in buildings. It differs from the static structure in that the services carry the lifeblood of the building, and each is often associated with major ongoing energy and running costs. These "living" systems need to be efficient, sustainable and safe, particularly in the case of specialist environments for healthcare, electronics manufacturing, and sensitive storage.

We have seen the advantages of BIM are greatest when the model is constructed early in the design process and that BIM encourages

this process. Building the virtual model early permits analysis of a range of design options to establish which is most cost-effective, while maintaining the most sustainable solutions.

Sharing the virtual design space means that MEP services engineers can resolve issues of complexity and clashes within the complete building framework and have an up-to-date understanding of the building, materials and situation.

It is obvious that the BIM model can represent the individual parts making up the services part of the building. In addition to the now normal benefits of coordination, clash detection, collaboration

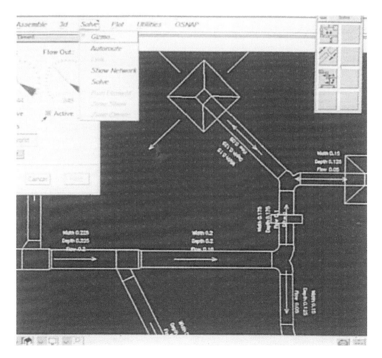

Figure 172. *REfLEX Solving network in-situ in model 1994*

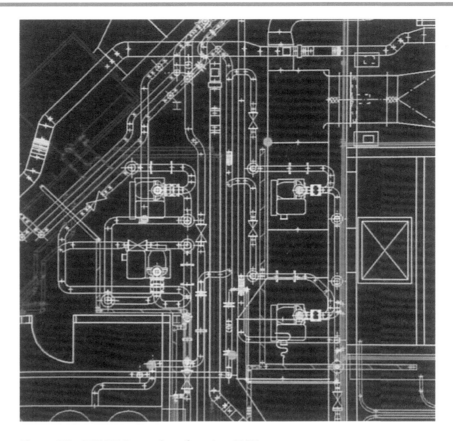

Figure 173. SONATA Screenshot of services 1987

and so on, BIM services brings a number of benefits particular to MEP.

The ability to form networks of objects that represent the particular services brings a powerful tool. SONATA and REfLEX both provided tools that allowed the design to happen with the model. Passing values between the joined ducts, pipes and electrical cables in the way that the windows communicate with the walls to say they are in the wall brings a whole new level of advantage. Duct sizes, voltages, currents, pressures, temperatures and so on can be determined within the model.

Additional benefits are that the network can be designed in-situ within the model. The automated design of duct networks within the model was achieved by having an extra view associated with services, where in the case of duct components included the flow of air, pressure, velocity and temperature. Sizing was done according to standards and tests were made to ensure connections between adjacent elements were correct. A physics view was also provided that passed information to connecting ducts. Friction

losses, temperature change, velocity, duct sizes were calculated according to physics and the standards. Figure 172 shows an example where by setting the diffusion rate into the room, the sizes, flow rates and temperature were passed back to the AHU. The room size determined the rate of flow of air into the room.

If the room changed in size, the system would iterate to find a correct solution to sizes all the way back to the AHU. The problem with the system at the time was the speed it took to iterate to a stable solution. In theory, changing the glazing in the room could affect the AHU in the same way and in theory the cabling for power. The Author never took it to that point but demonstrated that the network was self solving. In theory the same technique could be used to solve the balancing forces (bending moments and shear forces) in a structure within the model. Machine speeds have increased significantly since then, so such calculations would be viable.

Once designed in situ, clash detection and other tests could be run ensuring a services network that is complete and accurate.

Other possibilities that were constructed, at least in an experimental phase, provided CNC views where each piece of duct could produce the appropriate piece of CNC tape (or code, as it was). Another application they researched was testing of wired networks from within the model and thermal zones.

Kyle Steward (BAM Construct) Consulting Engineers used SONATA successfully for several decades. They had bought 30 workstations of SONATA for all disciplines in 1989, and subsequently developed a sophisticated ductwork library. This library integrated calculations, dimensions, joins, etc., into the different views, including shaping for the 3D duct and pipe pieces. With a coordinated set of services components, complex modelled parts gave them certainty and professional edge relating directly to manufactured items.

Several services libraries were produced (see Appendix 3 from 1988) and used to great effect, including the SONATA library and the highly sophisticated BIM library. The library included ductwork, pipework, electrical, drainage and schematics with 3D and symbolic views. The elements all included scale dependency (automatically changing the level of detail at different scales), full parametric sources, full clash detection and built in rule checking to ensure standards compliance, hidden line and link points. Link points enabled data to be passed from one element to the next, snapping to correction position, and variable passing when solving object networks.

Automatic network layout of the components making up services networks, such as pipes, cables and ducts, and automatic component sizing and joints between them within the BIM model creates significant savings. Tools for thermal analysis and system performance have been available since the first systems, ensuring that the buildings are as efficient and sustainable as possible. Obviously, budgeting of the services becomes very precise, to within a few percent, within the BIM model, differing entirely from traditional methods.

Life cycle management of the various components in the model can be easily achieved and is enhanced by accessing usage and performance data from the real environment. This might include operations and maintenance activities, ordering spare parts and shutdown and disassembly information. This would be achieved by having the component objects making up the model monitoring their real world counterpart with a live connection, similar to the coffer dam monitoring in the Heathrow Express project discussed earlier.

Live warning systems communicate from the actual building component into the model and performance data and operating conditions can all be drawn on for sophisticated analysis. Operational data can be compared with live data over the life of the building within the BIM. This leads to a better understanding of the components, their efficiency, and gives an understanding of how they function within the building, allowing refinement for future projects. Operational costs and sustainability can also be addressed with this data.

Building Services / Mechanical, Electrical and Plumbing engineers were the first users to take full advantage of coordination and parametric calculation capabilities within the BIM. Amongst the early users were Amey Mectric Ltd, British Gas, Haden Young, How Design and Management, How Engineering, Longstaff and Shaw, Robert Hayworth Group, Shepherd Engineering and others.

BAM Construct were still using SONATA for services work in 2010 and was still on their desktop after that in 2015. See the last page of this book! This probably gives Nick Crane and his colleagues at BAM Construct the title of the longest continuous user of BIM in the world. Figure 262 shows the Author borrowing Nick's computer running SONATA on a 2010 project.

CASE STUDY: BRITISH LIBRARY SVM

The British Library is the National Library of the United Kingdom and is the largest library in the world, with some 170 million items, some going back to 2000BC. The building was designed by Sir Colin St John Wilson at a cost of £500 million and was opened in June 1998.

Essential to the project were the MEP building services, designed by Steensen, Varming, Mulcahy (SVM). Their design embraced a wide range of services, including circulation for personnel and records, temperature and humidity control, security controls, lighting, sprinkler and other fire systems, plumbing and drainage systems. Half of the cost of the building lay in the provision of these services. "[Professor] *Colin St John Wilson and SVM.*" decided to use *"the same CAD system, SONATA, to the benefit of their client the PSA"*.

The most important service is environmental control, as one of the key roles of the building is to preserve the huge collection. Different areas, including closed-access book-stacks, readers' spaces, public areas and offices, require different conditions for both temperature and humidity. Book storage is located

Figure 174. *Photo of British Library basement 1998*

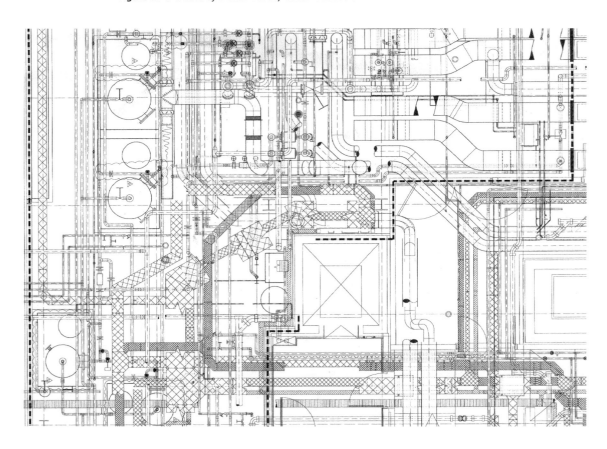

Figure 175. *SONATA drawing detail British Library "Boiler Room Coordinated Layout" © SVM UK*

Figure 176. *SONATA British Library Hidden Line Services © SVM UK*

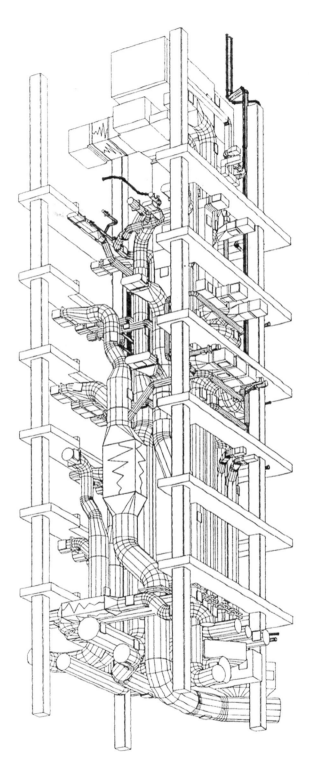

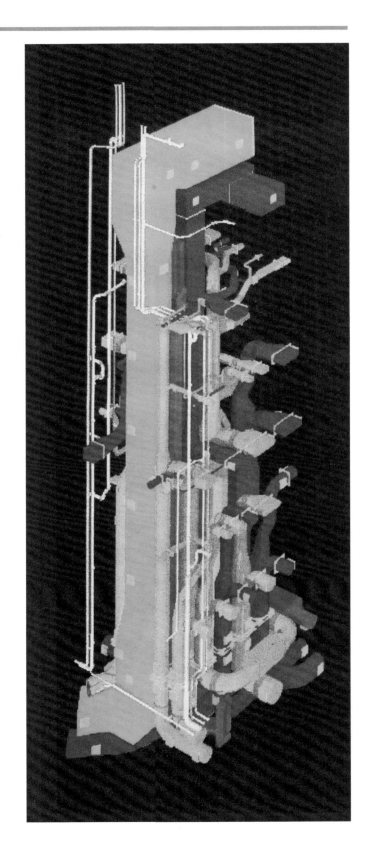

Figure 177. *SONATA model © SVM UK 2010*

Figure 178. SONATA model © BAM UK 2010

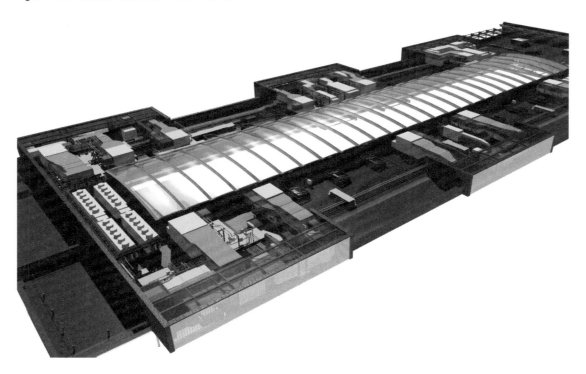

Figure 179. SONATA model © BAM UK 2010

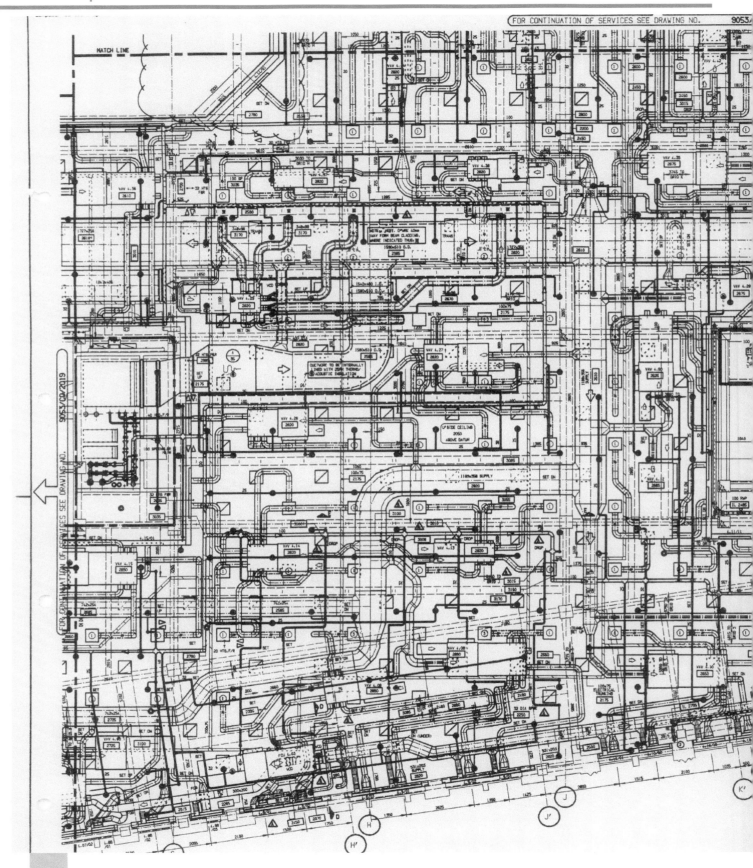

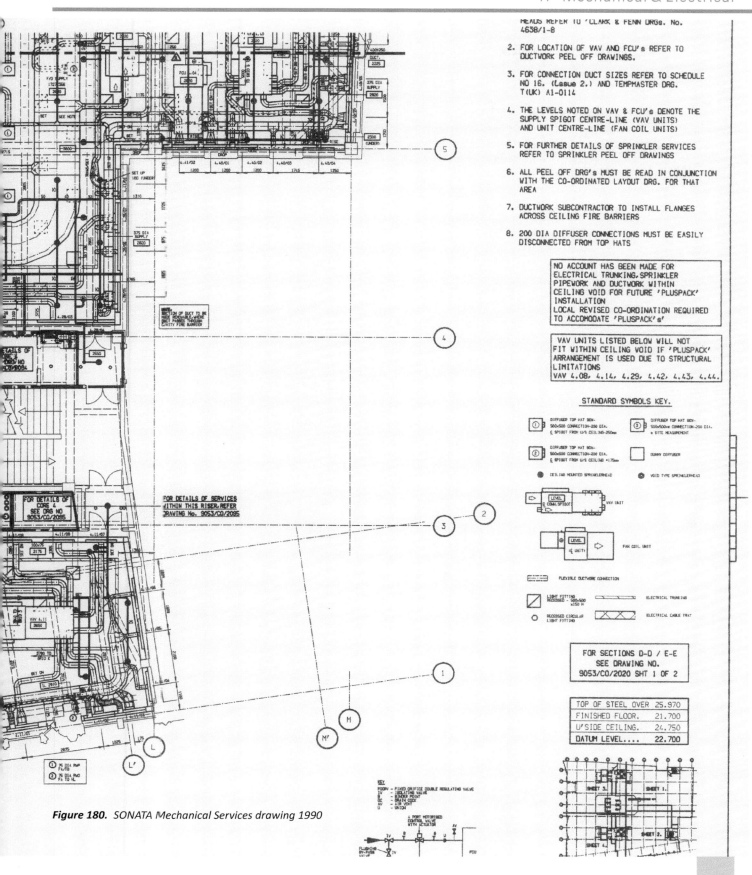

Figure 180. *SONATA Mechanical Services drawing 1990*

three levels below ground, secure and temperature stable. The external effects of temperature and humidity are almost eliminated, simplifying maintenance of the stable environment crucial to conservation.

A reinforced concrete secant pile wall surrounding the basement provides considerable mass, which taken together with the rest of the structure and the thousands of tons of books, provides an inertia against change in the environmental conditions, for which

air-conditioning does the rest. Fire-fighting was agreed only after much debate, as librarians hate the idea of water: a double action system with dry pipes being filled in an emergency and released only on confirmation of fire was finally agreed.

To quote Nick Crane of BAM Construct, UK, "*The last project we used it on for MEP, would you believe, only finished in 2010, as we couldn't find any software that was good enough to compete with SONATA!*"

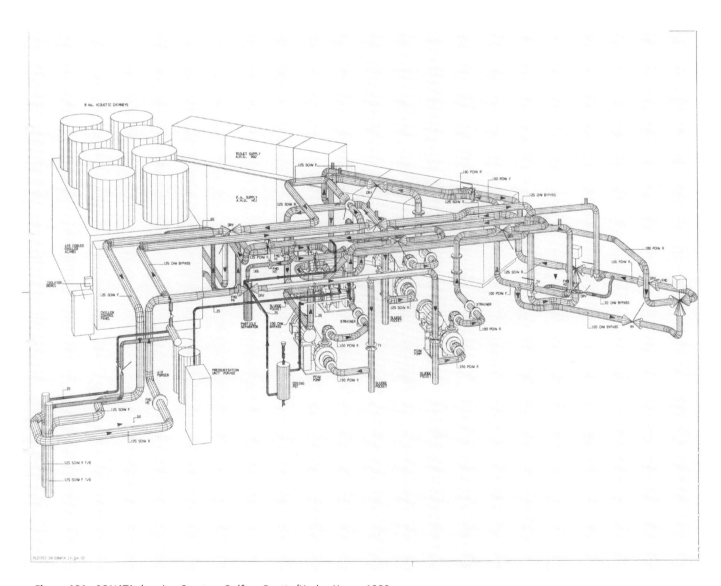

Figure 181. *SONATA drawing Courtesy Balfour Beatty/Haden Young 1989*

Evolution of BIM

Building Information Modelling is ever-evolving. Various applications have found their uses for the idea of Information Modelling, the concept behind BIM. Here we explore various developments of BIM, possible and actual, exploring the application of artificial intelligence, of new visualizing technologies such as AR, VR and GPUs. In these chapters the Author is seeking to bring these technologies back to BIM as much as it is to expand to new realms.

Retail Modelling – RIM   12

Retail Information Modelling (RIM) applies the principles of BIM to the Retail Industry. As in the Construction Industry, diverse information is managed, accessed and manipulated by a range of stakeholders. This information is not building fabric or construction detail, but rather the physical and data properties of products, planograms (arrangement of products on shelves), store fixtures, store layout, checking layout and related store items. Rather than reproduce the building information model in this structure, this model sits beside the BIM model, with direct synchronous access to the full Revit BIM model.

We show how a single source of intelligent information, together with appropriate tools for generating and maintaining that information, can help the retailer generate, operate, manage, and optimize their retail stores. It can also provide customers with detailed product information, Online shopping capabilities, and an understanding of customer needs. The store and the customer can also benefit from optimized layout and just in time products on shelves.

In this chapter, we discuss how BIM and RIM sit side by side and how new technologies assist RIM with design, intelligence, and capability. One might expect that a new BIM system will use some of the tools outlined. It would be exciting if one could manipulate a BIM model in Augmented Reality (AR), with hand gestures and voice commands, with an AI understanding what you were trying to build. This is what is being achieved in RIM currently. Also of interest is how the BIM model is integrated smoothly into the RIM model. There is no duplication but more of a smooth overlap. This technology is provided by Unity and Autodesk.

This chapter also gives insights as to how a new BIM systems might be developed. The uses of the technologies, development tools and synchronisty would play a part in any new system.

Acknowledgment: The RIM concept stems from a collaboration between the author and Mark Edwards. Edwards founded the 345 Holdings and VergeVT companies in Australia and this has led to the development of the system defined here. It is founded in part on the principles of BIM ideas and develops them in way that leads to a greater understanding of what is possible with BIM. This chapter is about the future of Information Modelling.

RETAIL MODEL OVERVIEW

As in BIM, information and structure are passed in and out of the RIM model with multiple disciplines sharing the same space. The result is a virtual store, thriving and dynamic, mirroring the actual store. This is called the Retail Information Model.

Multiple benefits accrue from such a model, improving management access and control, customer engagement, optimization of layouts and product selection, predictive behaviour of customers, and a range of other useful and potentially profit improving characteristics.

A range of tools has been created to deal with the diverse problems of the RIM model. Some of these are mobile device inputs that are used on the shop floor. The RIM product database has been extended to include more information about products, their origins, nutrition, GM, allergies, and so on. This information is collected from product labels, "scraping the Internet" and from the specialist image and capture companies. The data sources are diverse, and RIM allows the objects or components to access this information from the cloud and other databases, including EPOS, ordering information, weather, holidays, and most other things that can affect a store performance.

Figure 182. *Planogram visualization and within the context of an integrated BIM/RIM model © 345 Holdings 2019*

NEW TOOLS

There are technologies available to improve the man-machine interface. The traditional pull-down menu and icons have existed in CAD since the 1980s (SONATA being amoungst the first) and since then, the technology has not changed. Many modern systems have been over-engineered in that it can be difficult, even for the practiced user, to find the particular command needed from nested levels of pull-down menus or ribbons or the clusters of icons. Even Revit libraries don't have an index.

New tools, such as chatbots, Artificial Intelligence (AI), games engines, Virtual Reality (VR), Augmented Reality (AR), specialist

hardware in smartphones, and visual-spatial positioning systems, are available to assist in these tools. They are not necessarily new tools as some have been around for a long time, VR and AI in particular. VR was used in 1994 by Taylor Woodrow on the basement of the Albert Hall. The image shown is a modern view of a VRML file generated in the mid 1990s when VMRL was a standard format used to view in Virtual Reality. See Figure 165.

Integrating data between the existing systems is crucial and new tools are required to do this. A range of proprietary databases is currently used to design and hold information relating to the retail store. Spaceman by Nielsen and JDA by JDA Software are used to store product placement information. Other databases deal with

Figure 183. *Store fixtures by department with merchandising capacity analysis © 345 Holdings 2019*

Figure 184. *Adjusting category flow within the virtual store © 345 Holdings 2019*

sales, orders, delivery, and so on. The RIM database integrates smoothly into these databases instead of copying the information, either replacing or referencing data from the external sources. In particular, Revit BIM structures can be accessed and synced, from within the RIM system. Data is not copied but rather referenced or in some cases, adjusted in the original BIM model, usually in the case of fixtures. Similarly, we allow DXF/DWG (a 2D drawing standard) input into several of the design apps to help with store layout when building the information model.

PLANOGRAM RECORDING AND COMPLIANCE

Planograms are predominantly a 2D graphical layout, sometimes with simple 3D modelling, of how products are laid out on the shelves in a store. Traditionally this information has been generated with pull-down menus and icons. Planograms are created to maximize revenue for stores. The planogram generation programs create maps that traditionally, shop assistants follow to place the product on the shelf. In some ways, it can be likened (at a stretch) to generating drawing details of say, reinforcing to be constructed on the building site. Both are prone to error. In the case of the store, an incorrectly laid out planogram means that sales information does not relate to the planogram in the database. This problem is prevalent throughout the retail industry, with compliance (the rate at which product shelf positions comply with database planograms) being as low as 60%.

Recently, Artificial Intelligence has been used to analyse photographs of the store and check their compliance in particular. A photograph of the shelves in a store is sent off for processing. AI is used to recognize products and determine if they are in the correct place. The problem with this is that often, perhaps 95% of the time, AI cannot or incorrectly determines the product on the shelf, invalidating the whole process.

The Author's approach to this problem was to use a smartphone and walk along the aisle and record the actual positions of the products on the shelves. The visual positioning of the modern smartphone along with feedback of where things are expected to be, provides an excellent way of finding and setting positions of products. There are a number of toolkits available to help do this, in particular Vuforia from Apple. This type of tool enables the camera to look, know where it is and to display any AR information over the viewed scene.

Additionally, AI is used to determine what the product MIGHT be, and is used to draw the determined product over the actual product. It is immediately clear whether the AI has the correct answer, and, if not, techniques are provided to correct this. See

Figure 185. *VR Store environments highlighting sell down on shelf and within the fresh department © 345 Holdings 2019*

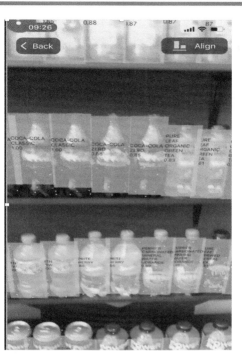

Figure 186. *Author's prototype with "Eric" giving in-store directions to products and facilities. Customer asks"Where is the Sultana Bran" © 345 Holdings 2019*

Figure 187. *AI product recognition and positioning from hand held device. Allows barcode scanning if AI unsure or incorrect. © 345 Holdings 2019*

Figure 188. *AR view of data generated instantly from Figure 187. Product information also available for research, database and compliance © 345 Holdings 2020*

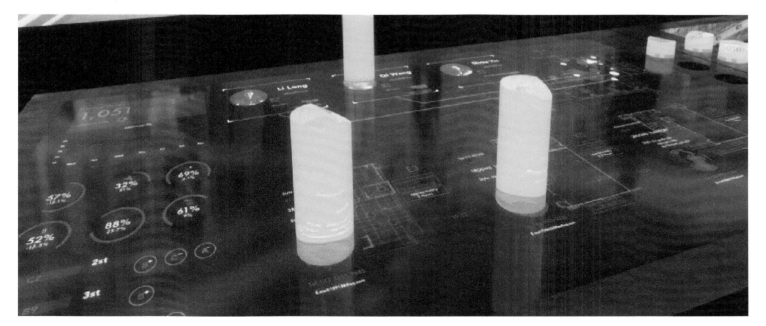

Figure 189. *Interactive data view from 345 VQ interactive Coffee Table © 345 Holdings 2019*

Figure 190. Typically the barcode on the product is used as this is unique. If the product is in the incorrect place, a warning appears in the AR view indicating where it should be. The smartphone/iPad refers directly to the cloud-based database for this information.

The same technique is used to generate planograms and store them in one of the databases, either JDA, Spaceman or RIM. We use an AI method called TensorFlow Lite to identify products on the device. Further information on AI is given elsewhere.

Planogram execution is the process of placing the products on shelves. The RIM approach to this problem is to provide a tool that hooks directly into the cloud based database, and via Augmented Reality, helps the employee to place the products at exactly the right place and correct orientation. The quantity of products, orientation and type are checked ensuring compliance with the original design layout.

PLANOGRAM GENERATION AND OPTIMIZATION

Planogram design is done traditionally using specialist programs such as Spaceman and JDA. These require trained technicians to operate them to place the products in the correct place.

To replace this, the Author has written a gesture control and speech-based system based on Google DialogFlow, and the Microsoft Hololens 2 AR system. This allows the user to manipulate the different products and layouts intuitively. DialogFlow allows the intent of complex commands in almost any language to be understood and acted upon. The Hololens 2 system allows the user to see the environment in which he is working, overlaid on the real world, and allows the user to grab and move objects, push them along the shelf and select different products.

Aisle design is done in the same way. Simple drag and drop in either VR or AR allow aisles to be arranged and rearranged in a few seconds, see Figures 182 and 184. Augmented Reality (AR) has

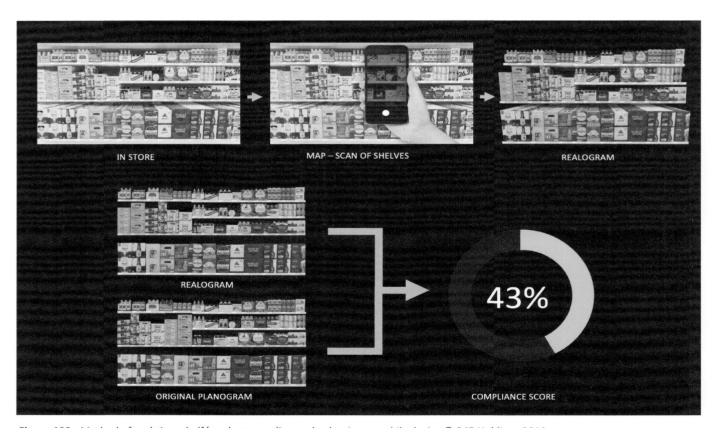

Figure 190. *Method of real-time shelf/product compliance check using a mobile device © 345 Holdings 2019*

the advantage over VR of being able to be used for long periods: VR is so intense it becomes tiring after a relatively short period.

So the retailer has various ways to generate planograms, scanning or assembling or in fact, using the traditional techniques of pull-down menus. The problem of creating an ideal planogram is more complicated.

Over the years, there has been a huge amount of sales data that has been generated. The problem that has arisen in trying to use this data to optimize store layout is that the actual layout of the planograms in store (as opposed to theoretical layout in the database) is not precisely known. Hence the sales data is based on an unknown layout or at best partially known. Here the EPOS sales data can be applied to the exact planogram layouts knowing that the store layout is guaranteed to comply with the planograms. Without compliance, the sales data is close to meaningless.

Artificial Intelligence enables the retailer or manufacturer to optimize the store, and planogram layout. Artificial Intelligence has the capability of recognizing patterns where humans see none. Compliant planograms with corresponding sales data create the possibility to predict the performance of any planogram placed in the store. Add to this capability of the AI to "scrape" the Internet for holidays, weather, special events, days of the week, advertisements, and so on, further improves the AI in predicting performance.

A later section discusses an app that allows an avatar (or arrows on the floor) to take the customer to where the desired products are in the store. This increases the correlation between a particular customer and their shopping habits and should increase the accuracy of the AI prediction capability.

The layout of the planograms and the store will soon be dealt with by the AI. Push the button to see how your planogram will perform, or push another to generate the ideal design.

RECORDING STORE LAYOUTS

Retailers and in fact, most building owners need to have accurate up to date information about their building layout. Some buildings have well maintained BIM models; most do not.

The retailer also needs this information. A particular client in the US has many thousands of stores but only has BIM models for a few hundred. AI is a tool needed to gather information about the existing stores and their layouts. This tool includes layout of the walls, using AI to recognise doors, windows and other fixtures. It is taught to recognize these objects using standard AI techniques.

Figure 191. *Real-time AI detection of doors, windows and fixtures © 345 Holdings 2019*

See the AI section later in this chapter for details. Additionally it is necessary for such a tool to sync with existing store models or drawings, whenever they exist.

Simple AR tools are provided to generate, place and position the wall and window objects on the smartphone to which ceiling heights can be added. This process generates an accurate model of the store from an internal perspective and is stored as an integral part of the RIM model. Planograms and their related shelves or fixtures are placed in the following way. The app allows the user to place products, POSM (point of sale material) and other information directly into the RIM model.

Figure 192. *Viewing total store performance in AR on the boardroom table © 345 Holdings 2019*

Figure 193. *Total store heatmaps highlighting customer behaviour © 345 Holdings 2019*

Figure 194. *Viewing planogram performance in AR © 345 Holdings 2019*

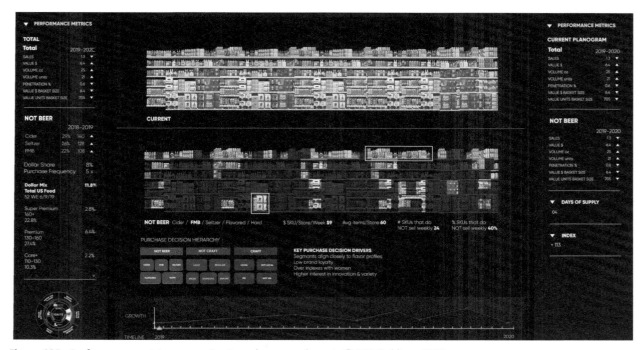

Figure 195. *Performance metrics comparing two planogram layouts © 345 Holdings 2019*

The Revit data is synced directly into the AR environment allowing the user to see objects in the BIM model live in the AR view. If the BIM model is changed in Revit, then the objects in augmented reality view also change in real-time. It would be possible to design around existing fixtures and arrangements in the physical space of those objects should the need arise.

Optionally the user can place this information into the RIM model or, as yet untested, in the BIM model. This uses the Reflect technology of Unity and Revit.

The same App can recognize products and their positions using AI and visual position techniques discussed earlier in this chapter. A planogram is defined uniquely by where products are on its shelves. The app can recognize the corresponding planogram it can see by comparing the recognized products on shelves with the database. This is done live. The app has access to this information via the cloud database and so can recognize from the pattern of products which planogram it can see, and exactly where it is. The planogram and the accompanying fixtures are placed automatically.

This actual copy of the store is known as a digital mirror copy. In addition to the standard maintenance and FM, this store can be used for Online retailing, providing a real-time model of the particular store in which the customer can shop.

VIRTUAL REALITY RESEARCH

Many manufacturers and retailers prefer to do their shopper research in a laboratory. This has been done in several ways, traditionally. Some of the larger retailers have very large physical mock-ups of their stores. The retailers send "customers" with a mission to do their weekly shop or shop for something in laboratory conditions. They are recorded, and after the event, questioned to understand their shopping behaviour. Obviously, this is expensive and time-consuming to arrange, and the problems of compliance exist even here.

Using a game engine (Unity), Virtual Reality (VR), and a database of planograms and store layouts enables the retailer to set up an interactive, photo-realistic, virtual store, to test the customer reactions and achieve highly accurate data of where and why the customer shops. The shelves are automatically populated by the products direct from the database (100% compliance), and the customer behaviour is recorded in a range of ways. This behaviour is analysed and questions asked via a number of graphical means. See Figures 185 and 197. Eye tracking, heart rate, and other

physiological characteristics are recorded tracking the customers reaction to the experience, the products and the store The Author produced the first prototypes of this in 2018.

PRODUCT DATABASE

The concept or RIM is extended over BIM to include the products and their details. Most retailers retain basic product information, manufacturer, usually the shape, size, images of the product, barcode, and a few other pieces of information. A broader range of data is necessary for a variety of reasons.

Customers are becoming increasingly aware of the products they are consuming.
- Nutrition, energy, and related information
- Warning customers of health hazards
- Guiding them on dietary suggestions
- Avoiding products from areas of conflict and high GHG costs[1]
- Minimizing packing and providing GHG costs for packaging
- Determining the recycling capabilities of the packaging
- Providing information about claims and endorsements
- Providing usage recommendations and warnings
- Minimizing carbon costs and meeting the new Government requirements of display greenhouse gas costs
- GM warnings
- Expected shelf life
- Preparation information
- Claims and endorsements
- Social media ratings
- Contact information

Management also needs a range of information associated with products such as delivery, shelf life, supplier, etc.

This provides information at the point of sale at the decisive moment of product choice. The customer engagement app discussed later in this chapter allows the customer to see which products meet their specifications and requirements as they are shopping.

Such information can be gathered individually by the retailers, but, like BIM objects with their increasing number of parameters, it is now available from some commercial enterprises.

---

1   For example, palm oil from certain regions has an immensely high GHG cost associated with clearing jungle for the palm oil trees.

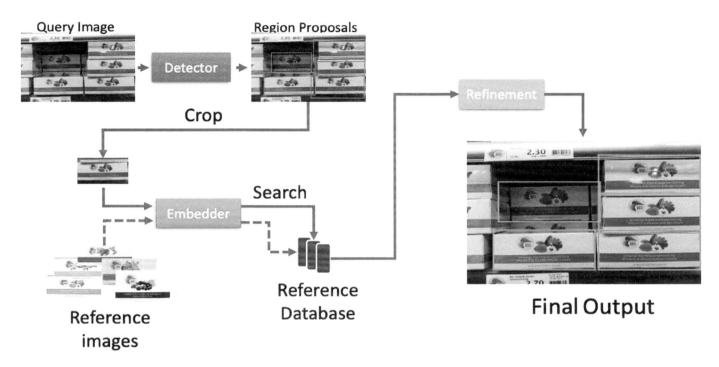

Figure 196. *Smartphone product recognition in real-time with AI and cloud database © 345 Holdings 2019*

As in BIM, different users require different information at different times. As in BIM, this information is available via the cloud, without being copied.

MANAGEMENT DATA

Management needs to monitor region, store, aisle, planogram, category, and product performance. A range of tools is provided to allow management to oversee all aspects of the stores.

This range of tools can be used using voice requests, in some cases hand gestures, in various environments including the board room, the virtual board room, dedicated spaces for viewing, on the show floor, and in web browsers. Figures 192 through 195 and 198 and 199 show examples of some of these.

Understanding the intent of an involved voice request is a problem that has been tackled by various companies, including Google with DialogFlow, Microsoft with the less known Luis, and IBM with Watson. This AI-based capability brings an automatic breadth of language and understanding and a simplicity of use that makes the whole process very simple for the end-user.

This, combined with the availability of the digital mirror store copy, the sales data, and other scraped data, an AI-based tool to gather and display the information and a range of hardware to display makes for a compelling data access and display system.

For example, management walks on the shop floor with AR glasses and sees the sales and related data projected over the actual products on the shelves. EPOS data can be tied to precise product positions and using new AI techniques, able to be tied to external information such as weather, events, holidays, and so on. Scraping data from the Internet is an essential feature of these systems and is intimately tied to product performance.

Management can also see the data at the board level, allowing access to multiple stores. RIM data is readily available, including such things as compliance for stores, category performance, and other significant issues.

RIM BUILDING BLOCKS

Here we discuss the capabilities and benefits of the RIM system. The actual detail of how this is achieved is outlined later in this chapter.

In this implementation of RIM we use parametric objects for a number of different purposes. In addition to the parametric fixtures, walls and windows, parametric objects have been used for monitoring stock levels of particular products and as independent temporal objects, passing information to stock ordering systems. In a way the store itself is one large parametric object with information being passed between apps as required, but stored only in a single unique position.

Another application of parametric objects within the store is passing sensor information and potential maintenance information to the store management system, as in BIM

STOCK TAKING

Stock taking is a time consuming and inaccurate process. Typically, store assistants go along each shelf, counting and recording the number of quantities of products. This data is then transferred to an inventory sheet, determining the inventory value divided over the different categories. We are automating this process to a large degree.

Combining a special camera (actually several cameras) with AI, the visual positioning system, and a link to the RIM database gives us a device that can determine the number of items on the shelf. As with any machine recognition system, it may not identify the correct products. This is solved by reading the product bar code. Using the AI probability features, we have an accurate view as to when the product count is accurate. As the AI learns, the solutions become more accurate. It will be able to estimate the number of sweets in a jar or pears in a box.

The ability of the camera to rapidly count the products, a shelf at a time at walking speed, means that the largest stores can be done quickly and accurately. If the planograms are compliant, the pauses for manual counting are minimal.

This information is stored directly into the Retail Information Model and gives an accurate count of most products. Counting loose products such as apples and, say, crisp packets is more problematic and some efforts are being made to solve this problem.

CUSTOMER ENGAGEMENT

The retail store is declining: by 2040 95% of all purchases are expected to be via eCommerce. In order to address this, new ways of engaging the customer must be found. The Retail Information

Model provides some tools that can be applied to engaging the customer, giving them new ways of refining and reviewing what they want or might want to buy.

The customer can now ask, of his smart phone;
"Where is the curry sauce"

in his local language and the response in his language might be
"it is three aisles over, let me take you there"

In addition, moving arrows appear on the floor in augmented reality and an avatar or not (of your choice) shows you the way. On the way to the curry sauce or the ingredients for Coq aux Vin, the system will show specials, chat about promotions and highlight anything that the (AI driven) system thinks that this customer might be interested in.

Similarly wide ranging questions can be answered such as "I need the bathroom", or "where are the cold beers" (as opposed to the warm ones) and "what coffee did I buy last time". Using information associated with the product definitions in RIM, customers have the ability to exclude all products with nuts or include all products with a low salt count. These appear as crosses on the products in augmented reality. You hold your phone up and you can see via the camera what is on the shelves and those products that are excluded have crosses over them.

Customers might ask for advice on a particular medication or ailment; "what can I take for my headache?" and so on: we are just scratching the surface of these capabilities. We use AI based chatbots (DialogFlow and language interpreters from Google) and bring our product data from a wide range of sources.

The ability to take the customer around the store, albeit to a particular place, and to show promotional items, or inform the customer on the way allows us to engage them directly never seen before.

The Author has produced a customer's shopping app, running on a smart phone, used to help the customer find items in store, to show specials, and simultaneously record research on customer habits. This app knows where it is at every moment. This is done visually, so no special hardware is required other than a smart phone or an iPad of some sort, perhaps attached to the shopping cart. The AI system tracks where it is and where all of the goods are relative to where it is. When the customer requests take me to the ATM, or where are the cereals, or I want to get all the things on my shopping list, the shopping app determines what it is and where it is. The best route to the items is determined, possibly with a deviation if directed by the store system. The route, the position, understanding of the language are all determined by separate AI systems.

This data is all setup by an employee walking the aisles with a smart phone running the Mirror App. This registers the products by position, image and shape recognition. This is again an AI technique. When not recognized, or the product is unavailable, the barcodes or labels are read. Using this data ensures the customer can find their way with the latest store layout and ensures best route for sales. It also enables comparisons to be made with the planograms stored in the database.

Once customers are using the shopping app, large amounts of research data becomes available. Every customer who pushes a trolley or carries a smart phone generates highly specific data that is recorded by the system. In order to apply a deep neural net to analysing the performance of a product, section, planogram or store, large amounts of detailed information is required. With the data generated by the shopping app, the actual performance of each of these can be determined. The actual performance of each can actually be determined, but the real benefit is having a scoring system for a planogram and store layout. This means that with enough data predictive performance of stores, planograms and layouts can be determined. AI brings the ability to analyse this very broad, complex data and find the patterns.

The shopping app provides this data in conjunction with anonymised till receipts (though linked by position and time) enabling predictive performance of stores.

MIRROR STORE SHOPPING

Using the various technologies, we have seen that one can generate a digital mirror store. One of the uses of such a store is to allow remote customers to visit the store Online and shop. Imagine using your iPad or desktop to go to your favorite store and see in great detail the store as you would see it in reality. You can select items, try them on your virtual self, and buy them. Shopping in Harrods, even if you are in China, becomes a click of a button. The guidance data and learned behaviour will all help this process.

The shopper also has all the guidance from previous section, and the information from the extended product database. Precise customer behaviour data is generated for every customer visiting. It would be possible to show the presence of other shoppers and allow conversations and interactions.

The technology is based on Google's Stadia, where the cloud engines compute the scene of the store and display it on your smart device. Stadia can present views of Saks on 5th Avenue at resolutions up to 4K and 60 frames per second with HDR and surround sound, meaning reality! It has dedicated lines for transmitting the images.

Such a setup enables virtual retail tourism, people from around the world will come to the store, to look and to shop. Gone are the costs of setting up the web pages for your store. Every store has an Online presence courtesy of the retail model.

ARTIFICIAL INTELLIGENCE

RIM uses sophisticated AI techniques at various levels through the RIM model. From the recognition of products through language understanding to learning customer behaviour, predicting behaviour, and to optimizing store, planograms and products themselves.

Here the detail of some of the AI techniques is given. It is an industry in its own right. All future design systems will have a heavy AI component and will eventually be controlled by AI. These next three paragraphs are complex, and unless you have a need, skipped.

The door, window, and fixture object detection of generating a RIM model from the smartphone uses MobileNet V2 followed by a Single Shot Detector (SSD). The SSD is a popular algorithm based on a feed-forward network. It produces a fixed-size collection of bounding boxes and scores for the presence of object class instances in those boxes. This is followed by a non-maximum suppression step to produce the final detections. MobileNet was created by Google and is based on a streamlined architecture that uses depth-wise separable convolutions to build lightweight deep neural networks. The combination of these networks provides an object detector which is fast and perfectly suited for mobile devices.

The neural network is written in TensorFlow, a machine learning framework written by Google, and trained on Google Cloud's AI Platform with a pre-trained model for initial checkpoints. The training process included hyperparameter tuning to increase the accuracy of the model. Before training, the input images went through an image augmentation process to increase the size of the training and test dataset.

Figure 197. VR Store environments © 345 Holdings 2019

A similar neural network will be used to identify fixtures and signs within the store. Although, when we perform object detection for products, due to the number of unique items, we have a two-staged process. A similar neural network to SSD MobileNet V2 is used to detect the type of product then an Embedder Network will discriminate the exact product from that category. The Embedder network maps the input image to a fixed-sized embedding via a triplet loss, bringing similar images closer. The embedded images can then be queried using the k-NN algorithm to find the most similar products. Even with the extra network, the detection can still be performed on the device.

One of the features planned is to have an AI control the complete store, from ordering to management, optimization, layout maintenance, and so on. This will involve not one AI but several (a parliament of AI's?) each controlling its own territory and in turn, passing information to the master AI.

LEARNING CUSTOMER BEHAVIOUR

Data about customers' buying habits and to some extent how a particular layout affects buying habits is often hidden and is extremely difficult to extract. AI can and will expose this data; AI excels at taking complex data and learning what it means. It can then score or rate new data depending on the very broad circumstances. For instance, an AI system could analyse customer buying patterns for a particular planogram and produce a scoring system for new planograms.

An example of a current AI system in the medical field is where, in the USA, patients have all of their case and family history entered into an AI system and the machine can predict, in a way that no person can, the likelihood of a schizophrenia appearing in that patient. It is impossible for even a trained doctor to tell why this is the case, the data is too complex and the signs too subtle for a

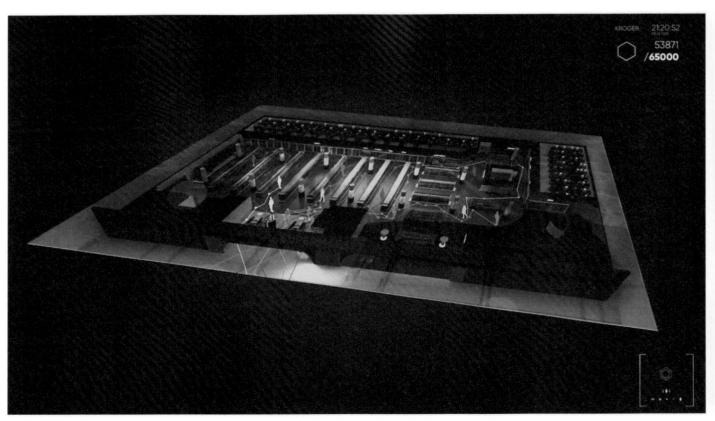

Figure 198. *Total store customer tracking © 345 Holdings 2019*

human to determine. But the machine can. What it cannot do is explain how it knows.

Using these techniques we can optimize product layouts in the store, the promotions, the pricing and the quantities displayed based on a vast range of criteria. The AI can scrape the Internet for related information such as holidays, advertising, weather and other external influences.

UNDERSTANDING NATURAL LANGUAGE

Rather than trying to understand and find one's way through complex pull-down menus, you can now ask questions, say what you are trying to achieve and the machine will respond appropriately either in an action or a verbal answer. This simplifies the use of what are becoming increasingly complex systems. Another use of AI that we currently use is understanding of the language. Given a command in writing (achieved by the Natural Language process of the previous paragraph), what does it mean? It is very simple to write a command driven system where precise words must be spoken. Learning about the structure of the language and understanding a range of language, perhaps to do the same thing requires AI. From the language comes the intent and the specifics of the intent. It is complex in that if one asks "where are the cold drinks?" and then asks "take me there" it is implied to take you to the drinks aisle. Currently our AI system knows 10,000 different products and can refer to them and find in different ways.

On top of this the command systems understand and speak many different languages covering perhaps 98% of the world's population. New AI techniques are becoming available to distinguish and hear a single voice in a room full of voices. These will be passed on as they become available.

Figure 199. *Total store customer tracking © 345 Holdings 2019*

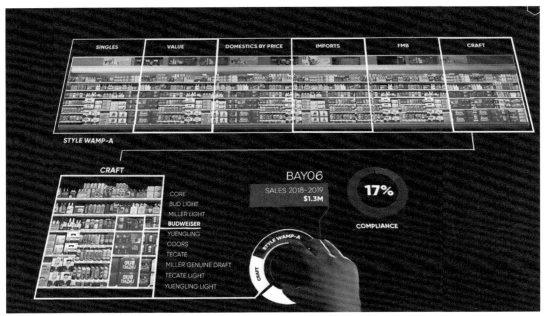

Figure 200. *Planogram flow iterations on the 345 VQ interactive 'Coffee Table' © 345 Holdings 2019*

CONCLUSION

Having a single Integrated Information model for all the store tightly tied to the BIM model brings a new range of benefits and functionality. Benefits include customer engagement, a mirror model accurately reflecting what is happening on the shop floor, an Online shoppable mirror store, accurate data available to view in terms of AR on the shop floor, tools to facilitate planogram placement and optimization.

For the first time the retailer is able to see an accurate model of exactly what is happening on the shop floor, and able to optimize and predict customer behaviour.

We believe that AI combined with other technologies such as Augmented Reality, Virtual Reality, cloud based data systems, dynamic virtual content and with the modern smart phone, brings a range of exciting tools for the customer, retailer, producer and researcher.

Figure 201. *Eye tracking heat map © 345 Holdings 2019*

Mirror Cities – CIM 13

Roughly 70% of people on Earth will live in cities by 2030, several billion more than today. This growth will place a massive strain on the cities and their infrastructure in terms of transport, energy, healthcare, water, and waste. To a large extent, urban infrastructure is already ageing and has grown piecemeal, further exacerbating these problems.

Extending the concept of Building Information Modelling (BIM) into City Information Modelling (CIM[1]) will bring the benefits into the design and layouts of cities. Extending the concept to a live model replicating the city, monitoring, maintaining, and mimicking should produce further benefits and capabilities.

Imagine a city, a virtual city, where you can wander, talk with the inhabitants, shop, check for information and visit as a tourist. Here transport, traffic jams and shops all buzz reflecting the life of the real city. You can immerse yourself in the virtual world, wander into a shop and buy as you would in the actual store. Information is available to manage the infrastructure of the city and the capability exists to operate the city from parts of the model.

What is proposed is a virtual mirror world, where trains and buses run and where people wander. This is an immersive world of the future. With an ageing population of 2 billion seniors in 2025, this technology will become essential for them and segments of the population.

Technology is now ripe for such a city. Internet data rates, the proliferation of BIM models of buildings, smart data from the utility companies, Internet data for transport, finances, and so on, the growing Internet of things (an estimated 20-26 Billion such things by 2025), GPS data from phones and potentially vehicles, VR and Enhanced Reality coming of age and of course Moore's

Law of ever-increasing machine speed means that this concept is increasingly viable.

Specialist tools are needed to deal with these issues, to ensure optimal development, sustainable operation and reducing costs. In this chapter, a technological solution is proposed to achieve this and to manage, monitor, design, maintain and optimize cities into the future.

This chapter presents a digital, temporal, information, BIM-like model of a city, combining buildings, populations, infrastructure, and transport. Information and control passes in and out of the model, multiple disciplines share the same space, and a multitude of viewers interact with all aspects of the virtual model. The result is a virtual city, thriving and dynamic, a mirror of the living city.

Potentially multiple benefits accrue from such a model, including improved design, accurate simulations, extended social and retail areas, social benefits, planning and monitoring, improved sustainability, energy, and traffic management.

The virtual city is not static but contains a "troupe" of object-based "actors", live, moving, and intercommunicating. These actors represent the dynamic parts of the model, from vehicles to lifts, from individuals to stressed beams. Data is piped into the model from a wide range of sources through the components that make up the model. In Chapter 10, "BIM and Engineering" describes Mott Macdonald in the Heathrow Express Coffer Dam project using REfLEX to bring displacement data into the model and display this as an exaggerated movement of components in the model.

Using this and related ideas, the model will run "in real-time" with live information coming in day and night. The model has an awareness of the "current time" or future time in the case of simulations. The model is viewed as a virtual city, with transport,

---

1   In this book CIM refers to City Information Modelling and Computer Integrated Manufacturing is referred to by its full name.

energy, people, building, maintenance, and endless other items mapping into the object-based information model. Individual objects such as vehicles or people can position themselves depending upon and change characteristics based on the incoming data or user interaction.

Currently, much information about cities is available on the Internet such as bus timetables, traffic flow, flight details, rubbish collection dates, and so on. Other information freely available includes weather, stock markets, opening hours, web pages, and so on, all of which can be made available in the city model. The list is endless.

In building a Mirror City, we create a model that reflects Reality. This would include things such as intelligent modelling transport systems, water and energy management and distribution, efficiency programs, traffic monitoring, road charging systems, smart parking, public information systems, consumption monitoring, security and emergency services management and control, management and servicing of equipment, crowd monitoring, CCTV control access and integrations and so on!

The idea of such an environment is not new. Gelertner's 1990 book "Mirror Worlds"[2] gives an insight into what one might expect.

Switches and functionality within the model can be used to control and interact with objects and data in the actual city. In 1996, REfLEX was used to operate light switches and doorbells in exactly this way. Data flows into the model, and operational instructions can flow out.

The aim is to identify and draw together into the Mirror City, all of this information. Adapting current BIM models into this environment will complete the city as a live, intelligent entity, that reflects Reality in every aspect. Let's explore the potential benefits to the city dwellers and planners.

The Author is fascinated by the idea of building a live virtual city and has planned for such a development.

MIRROR CITY OBJECTS

In this section, we examine the system engineering model required to make up the city.

There are various steps to complete a city database. It will be enormous and complicated, though not impossible because it would be cloud-based.

Separate BIM models would represent the buildings and the infrastructure. The Singaporean Government currently requires data from its building designers to be submitted to the City. If these submissions were extended to include a full BIM model, perhaps as part of the planning procedure, then the city would have an increasing number of buildings available. Extend this to infrastructure, and one has the beginnings of the model.

A system would be required to build and assemble the engineering part of this model. Similarly, a separate system would be needed to maintain the dynamics of this structure.

In simple terms, it would be a higher-level structure with building references and parameters for those buildings and access to them. As in BIM, the city will be viewed in different ways, and different information will be passed around. Much of a city is unmoving for long periods, and the model reflects this. For example, the buildings and infrastructure don't move. What does move are vehicles and planes, the people, money, power, water, and so on. The slab depth may change during the design, affecting other objects, but once the building is constructed, that slab is fixed. In the CIM, large numbers of objects are varying on a moment by moment basis. This model is no longer just about a moment in time, and this reflects conversations, vehicle and people movements, data display, power flow, and a host of other things. And it is available to all to view in their different ways.

Different types of objects make up the city model. Currently, we are used to parametric BIM objects that pass information to other BIM objects. In terms of buildings and infrastructure, this will not change. But there is at least one other class that acts differently and operates in time. This is not a recognized BIM object but a dynamic object that moves and interacts with other dynamic objects and as well as fixed BIM objects. It has a continuous flow of real-time information from potentially multiple sources. The dynamic objects populate the fixed city model.

The CIM maintains consistent information within the model in precisely the way that data is not duplicated and is shared within a regular BIM model.

The city model references external data such as Google Earth and similar programs. The dynamic information is something different. It is also different from the "I" in BIM. That information passes between BIM objects and is fixed in a temporal sense.

---

2   "Mirror Worlds", Gelertner, D., Oxford University Press, 1991

Another requirement for the database is for the city users to be able to build and populate their buildings. For instance, products into store, books into libraries and water into reservoirs (perhaps!). Avatars representing individuals should be able to shop this space. It ties the mirror store into the city in a spatial sense. Virtual coffee shops, cinemas, theatre and so on will come to life in the virtual city. Providing this capability brings a richness to the model and a new aspect to cities and their visitors.

It is suggested later in this chapter that individuals or their avatars will be able to move freely in the CIM to see and to be seen. They will able to enter these virtual "spaces," looking at the world and seeing goods, information, and so on, as one would in the store or perhaps a library.

The dynamic objects themselves control the flow of information into the model. The objects themselves control the display of information depending upon the viewer and the external data. (as in BIM). Figure 203 gives an overview of the CIM in terms of information flow. This range of data is achieved through different types of connections, mostly, but not exclusively Internet-based. The objects making up the CIM look after their own data inputs, objects make their own connections with the outside world. Your personal avatar may (or if you choose, may not) be attached to a character in the mode. As you move around the city either virtually or actually, your avatar moves in the virtual city. In an engineering sense, a concrete beam in a bridge might have a fibre optic sensor attached to it measuring some aspect of performance of the beam, perhaps stress or strain. That information is read by the beam object and could display as warning or data to the viewing engineer.

For critical applications, the data might be streamed directly into the model, whereas for long term monitoring, the data would probably be sent to an intermediate file and accessed at intervals by the model. The actual code to access and display the data is built into the beam component as part of a component "view" of the beam. The general rule is that if an ordinary computer program or app can access the information then, with the appropriate library linked into the overall program code for displaying the CIM, the information becomes accessible to users of the CIM.

As with information model systems, each of the component views are assembled (cross-compiled) into machine code and made into Dynamic Linked Libraries (DLL), a process usually made invisible to the user. These DLLs are distributed over different machines across the Cloud. For instance, a sensor at a traffic light at a

particular street intersection may be defined on a machine at the intersection and linked back to the appropriate CIM object.

Active components will no longer represent just building elements, but also; their physical state, their position, their attributes, and may themselves be display boards of information, ticker tape and CCTV. They could present themselves differently via the various view mechanisms to different outside parties.

Now we examine the types and sources of data, how they might be viewed, and how these data would provide a useful addition to the virtual city model.

PEOPLE, NPCS, AND AVATARS

One of the powerful capabilities of the CIM is that it is possible to have moving objects inside the model. The Author did similar work in SONATA where a man (sitting at a desk) moved around the BIM. He was always there, no matter how you looked at the model. It was an ordinary component that moved under its own volition. He moved slowly and through walls.

With modern smartphones, it is possible to track the position of the phone accurately. A cellular phone company has completed a study where it tracked all people entering an area of Hyde Park in London and mapped movements accurately. Using this information and other techniques such as WiFi mapping and visual recognition, more accurate placement of individuals can be achieved. Modern cell phones also have the capability to track their own positions visually. This is done by recognizing objects and tracking sensor reaction to motion. Examples of this technology (used in UnderstandingBIM app) are ARKit and ARCore. See Retail Information Modelling Chapter.

People and vehicles may choose not to be displayed although cameras with AI analysis will always know where there is a car, though not a particular car. Similarly the information displayed may be degraded, depending upon the viewer, maintaining privacy in the same way that GPS signals are degraded depending upon the viewer. The conditions and the degree of degradation, or even access, is very important and would depend on many factors. The security of the city in the CIM would be critical.

General positional information and crowds are known from other sensor information. It is proposed to populate the CIM with NPCs representing the non-specific people (those without phones or those that choose not to be present), crowd and vehicle densities. In computer games, an NPC is a Non-Player Character, a virtual person who is not fronting an actual person but is controlled solely

LCD WALL SCREEN

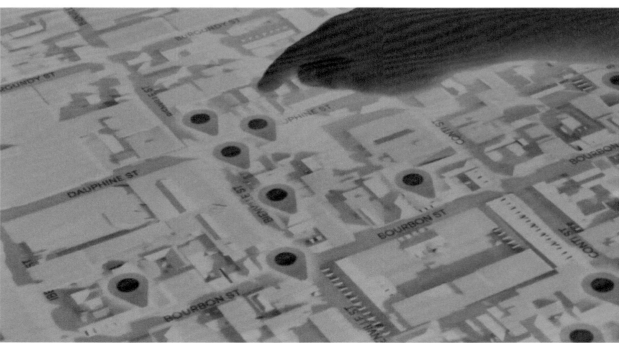

Figure 202. *Retailer Interacting with City using touch sensitive screens and AR © 345 Holding 2020*

by a program. They achieve different degrees of intelligence in the games to the degree that in some, it is not possible to differentiate between the NPC and real character. Computer games allow voice or text communication between the different characters and it is planned to make the avatars and the NPC communicate in this way. Currently the Author is using Unity to model the different environments; verbal and text communication is standard.

The benefits of having real people inhabit the virtual city would be many and varied.

PUBLIC HEALTH AND SAFETY

One is not limited to knowing just where people are in the model. Sensors are available to monitor condition. Vulnerable people might be tracked closely. In vehicles, one can monitor oil pressure, tyre pressure and temperature, tread wear and speed. This information could be used to avoid heart attacks, blowouts in trucks, and so on. The CIM becomes a way of displaying Big Data.

Apps and systems exist to monitor basic information about an individual. These are recorded directly into the smartphone or via WiFi or Bluetooth, and so are available in the cloud. Such sensors might monitor physical body properties, such as glucose, blood pressure, and temperature, with a tiny battery-less sensor placed sub-cutaneously, allowing their phone or a system to read off health as they moved past certain readers. The position and the condition of those at risk would be flagged to a view into the CIM.

Monitoring public and individual's health could be part of a CIM system. 30% of the western population is expected to have diabetes by 2030, a real reason for supporting such technologies. The largest killers today are heart attacks and strokes. Attaching an AI to monitor conditions of the elderly or those known to be at risk would potentially predict when a heart attack or stroke is about to happen. Given that the system can locate the potential patient, this has obvious benefits in terms of prevention and rescue.

Monitoring health in this way might lead to new means of mapping illness, diseases, and help determine relationships between physical effects, zones, buildings, and circulation areas. This can happen now, using smartphones, but with a monitored population, AI will learn faster, and health authorities will be able to assist.

Disaster control, dealing with flooding, evacuation, terrorist attack, disease monitoring and control, crowd control, or rioting would benefit knowing numbers. AI systems exist to identify crowds and their sizes and potentially individuals. Bringing crowd situations,

and disaster situations live into the model would allow emergency services to manage these situations.

For instance, it was an AI system run by a company called BlueDot, that first spotted the Covid-19 virus This company uses machine learning to monitor outbreaks of infectious diseases around the world and noticed an unusual bump in pneumonia cases in Wuhan, China. The Wuhan CIM model with appropriate AI (if it existed), would flag such occurrences automatically to other CIM models giving all an instantaneous warning to all.

The modern problem of loneliness could also be addressed. For the elderly and infirm, being able to wander through the city as an avatar, being able to talk with other "people" without having to leave their living rooms could be a huge boon.

Taking holidays takes on a whole new meaning, something akin to the Schwarzenegger movie "Total Recall."

WATER AND WASTE MANAGEMENT

Experts suggest that one of the major problems of the next 100 years in cities will be the lack of water, making its management a critical aspect of smart cities. Meaningful and actionable data is collected about the flow, pressure, and distribution of the city's water. This is known as Smart Water.

Modelling the pipes, the treatment plants, the catchment areas, the pump stations and monitoring them, bringing information in a single unified model would have wide-ranging benefits. Sharing of information between the different departments involved in this process would bring BIM like benefits, For instance, the watershed management team can share their "view" of the CIM to those concerned with flooding, to those working in wet areas and any impact on the transport would be apparent. Monitoring this data might be live and could become predictive with precipitation intelligence.

Another example is monitoring water loss from the supply system. Knowing where the older pipes are, where it has been frosty, what the flows and pressures are at different points are and should be, where engineering works are carried out all would assist in pinpointing losses. Control of the networks from within the model would also be possible. The ability to control valves, activation of equipment in purification, treatment, or pump plants, closing sluice gates could all be controlled from within the model. Warnings of different types, fluid levels, and energy usage might also be monitored and regulated by the relevant components in the CIM.

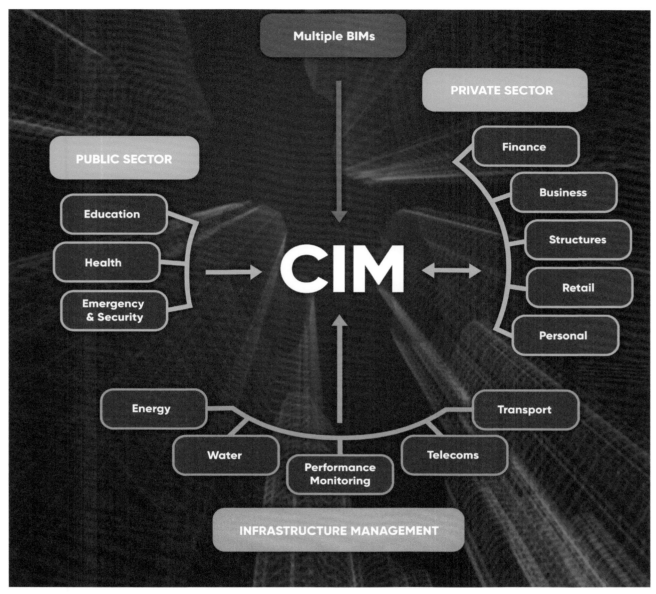

Figure 203. *Guide to CIM model with inputs and outputs.*

Smart water meters are becoming more common. This information could provide valuable input to home and water management. Measuring figures on-line would mean they could be displayed and built into the model. Usage, integration of data from different systems, integration with run-off data, rainfall, predictive Met Office data, and so on, would lead to an integrated water model.

Pipes would be tagged with sensors, and flow rates measured at critical junctions. Connecting the pipe network to the CIM would enable monitoring and active management of water.

With sensors across the system feeding back into a unified model, waste and potential problems can be addressed at an early stage. An awareness of the overall system performance, potential maintenance, and problems can be drawn together. Again an AI to monitor and manage the system would be relatively straightforward once the complete system is modelled.

TRANSPORT

Transportation systems play a vital role in dense urban environments. Billions of people need to get to their work on time and safely every working day. The management, maintenance, and control of the various transport systems are crucial to the economic and social fabric of all cities.

Much data is available from current sensors and recordings. Information about exact positions and arrival times of planes, buses, trains, and subways is known. Traffic light monitoring, traffic monitoring with AI, and CCTV provide extra information. Linking RFID ticket reading machines such as in Seoul and London allows systems to track patterns and individuals.

Cars and trucks are different again; as the position of individual cars is not generally known at any moment, there are means to track them. Information from number plate recognition cameras, congestion charging cameras, AI tracking systems, in-road sensors, and smartphone GPS information can all be used to track vehicles. Data concerning how many and which cars going through an intersection can be measured with software alone. For example, London Transport Congestion Zone Charging already knows how many, how fast and potentially who is travelling.

Technology exists that measures different physical properties of sensors, where the sensors are battery-less, penny cheap, permanent, and identifiable. A single capacitive sensor added to an RFID chip circuit to modify the frequency of transmission proportional to the change in the property measured. No battery is required. The Author patented and constructed this device for use with truck tyres. A variant on this measures the tread depth. The ID, pressure, temperature, and tread depth can be monitored in the truck or by sensors built into the road. Figure 206 shows the tyre pressure being read. This was built and implemented by the Author at a truck depot in Cambridgeshire, UK. Each RFID chip has a unique identity number. This information fed into the model could provide road usage, safety, legality, tracking, security, speed, and other information for the different parties.

Traffic flow and traffics lights can be managed more effectively with a single source of information using AI. Emergency response vehicles would have access to the CIM model and would be able to manage lanes and lights accordingly. Sensors can be embedded in buildings and infrastructure and monitored on an occasional basis. They could measure stress, strain, humidity and corrosion. They should last the life of the building as the power source comes from a magnetic field.

The Department of Computer Science and the University of Illinois has conducted research where vehicle movements have been monitored in the form of time, location, heading, and speed as

Figure 204. *SONATA Image City of Melbourne BIM Model, 1988, by Jeff Findlay © Building Modelling Company*

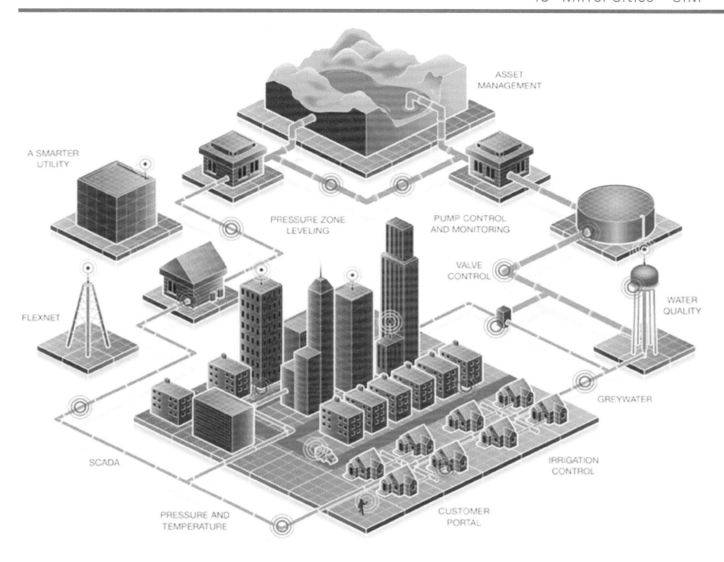

Figure 205. *Smart water network. Image © Sensus Inc., www.sensus.com 2015*

often as every second. This data was collected by Nokia, Verizon and other companies through the navigation provider Navteq. The goal of this project was to simulate vehicles and their behaviour with input from real probe vehicles. Problems of accuracy and sparseness of data will become less problematic. They make reference to obtaining further data from sensors in the roads and RFID.

The public view of the CIM would provide what is happening and where bottlenecks are appearing. Some "added AI" would give a view as to where the bottlenecks will happen.
Currently apps track buses, planes, trains, and subways, adding this information into a model could enable the public to see

exactly what is available and what is happening. For instance, Flightradar24, an app for smartphones, provides an exact position and other information about planes already.

The different vehicle types would be dynamic objects in their own right, that move through the model, each pulling its information from the Internet. Crowding, connections, potential crowding predicted by AI, and a host of other benefits would result. The virtual instance of the train would change as the position of the actual train changed. If one were to take a snapshot view of the model, all the trains and planes would be exactly where they are in reality.

Coordinating the transport systems together with other municipal activities, maintenance, events, and emergency operations would enable a better-controlled city.

EXTENSIONS TO BIM

There are various technical problems that must be addressed to create a dynamic city model.

It is unlikely that there will be a complete information model set for the city. Various systems exist for generating non-information-based models, including tools such as Blender, 3D Basemap, Google Earth, and Google Maps. Basemap, in particular, contains data at different levels of detail, necessary for an efficient viewing mechanism.

There will be a vast amount of information coming into the city model from many different sources. This can be managed in a distributed, cloud-based structure.

Gaming technology currently deals with precisely this problem, some having over 500 million registered players and millions online at any moment. Game technology (Unity) was used for most of the systems in Chapter 12 Retail Information Modelling. As a game, much of our city model is static, and this could be pre-prepared for efficiency. The original models are held intact and if changed, will flag up as altered. The large user base of "players" will not see these models but rather the processed models spread amongst the servers. Town planners might need access to the live BIM models for certain types of data and so, in their smaller numbers, could achieve this.

Games Engines provide a cross-platform environment (runs on different types of devices and machines) that copes with the multi-user, high-quality graphics in real-time that is needed to run a realistic world for the users.

We need to include NPCs in this world. An NPC is a Non-Player Character. Normally these are zombies or equivalent that wander around the game, driven by an AI engine, trying to kill the gamer. Here we are changing the character of the zombie into a bus or train, or perhaps a car. These vehicles are driven by data from the real world rather than an AI. Other objects are fixed but still have animated characteristics. In the 1980s, the Author and colleagues built cranes that could be operated in the BIM. This can be done fairly quickly if a programming environment is allowed for particular views of an object. That function representing the particular view of the object refers directly to the data stream, and when the data changes significantly, the object flags itself to be redrawn. Moving objects also operate this way.

The avatars representing individuals in the scene are also represented by the objects, they move, using animation sequences defined in Unity or equivalent, and pass information backwards and forwards about position, speech etc.

Required Changes in the parametric language include;

- Additional functions to adjust its position, rotation, and animation sequences, access external information live from the INTERNET
- Flag when it needs to be redrawn by the display system
- Access to microphones, speakers and INTERNET voice streams

The different objects that make up the CIM should have potentially many extra "views". These are separate definitions from the standards graphics views of elevation, plan and 3D, perhaps a view or every type of user who might want to see this object differently. An example in the early systems was the "physics view" in SONATA and REfLEX that was executed repeatedly in solving networks of the objects. It was used to propagate information around the duct network and was seen by other physics views. Another example was the dashboard that operated the crane in these early systems; it was seen only by the crane operator.

The concept of zones would need to be extended to allow a range of areas within the model to be defined. These might have soft or fuzzy boundaries and may be dynamic in their own sense. For example, some of these bounded areas might include a neighbourhood in a social sense, proximity, region, zone, crash likelihood economic zones, shadow calculations from daylight blocking and so on.

Various others and commands and functions need to be added to achieve a temporal aspect to the system. These might include time to, the time elapsed, follow the path, follow the road, flag, end error, send a warning to email, follow without crashing, update at time intervals, what date-time, simulated date-time, update every frame and so on.

For instance, an elevator or lift will have a maintenance view that, several weeks before maintenance is due, will connect with the relevant people. It operates in the model in terms of moving between floors via functions in a general view. The actual history of loading and operation may be maintained and the amount of current drawn to operate will be either in the component itself or a connected object. A CCTV camera inside the elevator could display in the model. It may be possible to operate the elevator from the model when an emergency occurs.

Additional view information may attached in the CIM associated with sensors and maintenance. Continuous data streamed from a fibre optic link can be fed into a file and analyzed by the beam component at intervals, or if truly critical, fed directly into the live model.

VIEWING CIM

Potentially, there are millions of people (and possibly AI's and other systems) trying to access the Mirror City model in different ways. Apart from the sheer numbers, there is the problem that different groups of the "viewers" will want to see the model in particular ways.

The BIM equivalent, for example, is that the service engineers want to see the model differently from the city planner, or perhaps the interior designer wants to see something different from the quantity surveyor. BIM allows for this in a simple way.

Different access modes need to be provided for different types of access. These modes might access particular views of the objects making up the CIM.

The tourist will want a different view of the city than say the Civic Engineer. These views will be provided with these modes, the mode defining the views that will be used to access the CIM objects. Another more diverse example is that the engineer in charge of water for the city will want a schematic view of the water pipes, while the driver entering the city might want to see the traffic situation.

The tourist might want to take the underground to 5th Avenue in NY and then wander the streets. The user could access this via VR and be immersed in the scene, with the other travellers, or perhaps on a web page. Again one of the game engines can easily achieve this diversity. Google's Stadia removes the various technical issues associated with remote real-time VR graphics.

Using a game engine (a game engine is a system that provides tools to generate games easily) such as Unity or Unreal Engine to create the Viewing Engine brings immediate access to a wide range of technologies and devices. This includes AR, XR, VR and the latest glasses technologies. These engines also bring the multi-user capability as discussed. The tourist might also like to take the superman view of the city or visit MOMA or the fun park. The RIM chapter describes in the customer engagement section, an avatar that knows the store and where everything is. Here he or she (or it?) appears to show the tourist around his points of

Figure 206. *Drive-over aerials measuring pressure, tread-depth, temperature, and ownership*

interest. The technology of visual positioning (used in RIM) solves the positioning problems.

If the tourist chooses to visit the city in person, then the AR display becomes ideal. You look to see where the next bus is, you can see it through the buildings or from a plan view. AR directions are appearing in some of the more popular map Apps. One might extend the idea of city tourism to viewing ancient models of that city overlaid in AR or in the pure virtual world.

Accessing data within the CIM should be intuitive. One would move through the CIM and ask questions of the what is visible rather than just a flat world of web sites. For instance, approaching a bus shelter one might click on the timetable to view details or actual see where the buses are currently.

Enhanced reality takes data or images and overlays it on a live camera view of whatever you can see. In the CIM, this might be done with special glasses, projections on to windscreens, or as shown, an addition to a smartphone or tablet. The latest AR glasses, almost indistinguishable from ordinary glasses, will be an ideal way to view this information.

Increased data rates, incredible CGI, fast games engines ensures that the images of buildings will be indistinguishable from reality. Some images claiming to be CGI are so realistic one cannot tell (see Figures 217 through 219). History has shown that the best images of today become real-time images of tomorrow. Rendering hardware found on most PCs and some smart phones transforms data into ever increasingly realistic real-time moving environments. Applying this arriving technology to our CIM could produce a very realistic world.

Figure 207. *Hololens 2 Augmented Reality view of a City Courtesy of Microsoft*

In theory one can take your avatar on the train on the virtual CIM; or perhaps wander the streets and see the traffic, the lights, operate the lifts and, if your lift is stuck between floors, remain there until it is fixed, or of course, just abort out of the program.

CIM would be a virtual world that mirrors the real world. Several movies have explored variants on the theme "let's stay in bed and live our lives through an avatar." This is slightly different in that one would have a virtual existence in a virtual world, something closer to perhaps the "Matrix." To make it work fully, people in the real world would need to allow their phones to state their position. This would give them a virtual presence; global positioning, body and voice would probably do it. One might even dress the virtual presence before one leaves the house in the morning. As one drives or takes the bus to work, your virtual presence would do the same. For those of us who are a little lazier, we may prefer to stay in bed and take the virtual bus anyway, or perhaps we would just beam ourselves up by using one of the other mechanisms. The concept of virtual tourism becomes a reality, saving people round the world flights and expense. There would, of course, be savings to the environment.

It will fulfill multiple purposes and will provide economic, social and ecological benefits. The technology to achieve much of this is available now and with access to BIM models and some development, a fantastical useful new world will appear.

BIM and Manufacturing   14

IMPROVING EFFICIENCY WITHIN THE CONSTRUCTION INDUSTRY

Global construction will grow $17 trillion in 2017 and is expected to exceed $24 trillion in 2021.[1] Unfortunately, the industry wastes more than $120 billion annually in the United States alone, because unforeseen delays, cost overruns, and other inefficiencies. The AEC industry could improve its waste record by following the lead of the manufacturing industries and by leveraging design collaboration, prefabrication and modularisation. It has been attempting to move towards the industrialisation of construction, by refining the end-to-end design and updating the construction process, but little progress

has been made. This chapter discusses some of these and other possible benefits.

It is quite possible that the ancient Egyptians saw the benefits of and used the principles of standardised work packets to cut stone, transport it to and assemble the pyramids. They invented a wide range of technologies to assist in pyramid building. This included the ramp, the lever, the lathe, stone drilling, the saw, some have suggested steam power, proportional scale drawings and so prefabrication of the blocks of stone, just in time delivery of those blocks and other techniques would not be difficult to imagine.

Somewhat more recently, the Crystal Palace of London, built for the Great Exhibition of 1850, was amongst the first buildings to be constructed with a modular system and work packets. The

---

1   "Global Construction 2030: A global forecast for the construction industry to 2030", PWC

Figure 208. *Crystal Palace, London 1854*

designer, Joseph Paxton, had experimented with glasshouse construction and had developed novel techniques for modular construction, using standard-size glass sheets, laminated wood and prefabricated cast iron. Each module was "identical, fully prefabricated, self-supporting and fast and easy to erect. The modules were easily constructed, each module being a work packet. This design cost somewhat less than 1/3 of the next contender and was considerably larger.

The modern theory of modernisation and standardisation started when the influential architect Le Corbusier visited Henry Ford's car factory in the 1930s. After seeing factories and his visiting several American cities, he launched the concept of the second machine age. Having been inspired by Ford's factory, Le Corbusier wanted to raise efficiency in the Construction Industry by applying industrial methods, mechanisation and standardisation. Several projects of the time, including Le Corbusier's "Radiant City", used prefabricated units and simple assembly.

In the 1960s, "systems building" evolved using standardised, prefabricated building components, with quality control, construction sequencing and new methods of documentation. The advance of systems building has possibly been hindered by the perceived ugliness of the resulting structures. Most buildings are still unique, each designed and built with little formal reuse of knowledge and experience, made to order by artisans working onsite and usually under difficult conditions. This is not strictly true for buildings in the Eastern Bloc.

A benefit of BIM is that it allows the construction industry to follow the other industries and to become better aligned with the manufacturing processes. The automobile manufacturing business has provided a direct model and is used to contrast and search for possible improvements in the assembly process.

AUTOMOBILE INDUSTRY

Following in the footsteps of the ancient Egyptians, Paxton and Le Corbusier and with the benefit of hindsight, we examine the Automobile and Construction Industry processes with BIM design in mind.

The automobile industry moved from craft-based production to global mass production with Henry Ford's innovations 100 years ago. In 1921, Ford was producing a million cars per year. He utilized (originally used in 1104 in the Venetian Arsenal) the idea of the moving assembly line reducing the T-Model Ford assembly time from 12 hours to 2 hours and 30 minutes. He introduced a 40-hour week and a good salary level keeping his employees loyal.

Figure 209. *T Model Ford Production Line, Detroit 1913 By Ford*

At the end of WWII, Toyota pioneered lean production, with which Japan rose to pre-eminence with automobiles in the 1980s. Some of the improvements in the process can be attributed to materials and technology. Many of the gains were made with the organisation and management of production. Even with modern cars that continue to be custom made to order, all processes have improved.

COMPUTER INTEGRATED MANUFACTURING

Computer Integrated Manufacturing[2] uses computers to control the entire production process. This allows the separate processes to pass information with each other and to initiate actions. Similarly, control and information passing are inherent parts of the BIM process.

Buildings and modern cars are complex with no two being identical; electronics cabling, variations and options apply to both. Different parties design different parts and everything must integrate in a seamless, reliable way. Computer Integrated Manufacturing uses information technology to solve these problems and includes *"the integration of various processes and people within the design and manufacturing process"* of the

---

2   Because of the acronym conflict between City Information Modelling and Computer Integrated Manufacturing the Author has chosen not to use the acronym for Computer Integrated Manufacturing but rather spell it out in full.

building, and "*It includes integration of business planning, demand forecasting and total management*"[3].

Rembold expands the concept of Computer Integrated Manufacturing to include the activities of CAD, Computer-Aided Planning, Computer-Aided Manufacturing, computer-aided quality control, Production Planning and Control (PPC) into one single system. Computer-aided quality control can tie the manufacture of building components by BIM objects, and production planning and control are covered in different aspects of the BIM model. PPC is the technique for forecasting every step of the production process and might tie in through intelligent time-based objects. Systems integration is also a fundamental part of the BIM design process. Building Information Modelling and Computer Integrated Manufacturing go hand in hand as they both integrate the entire design/build process. Both use automated design processes with Computer Integrated Manufacturing probably being superior on the automated assembly or manufacturing side.

Computer Integrated Manufacturing uses closed-loop control processes to automate manufacturing. The aim of Computer Integrated Manufacturing is, in part, to completely automate the manufacturing process. BIM has yet to use closed-loop control processes in automation of the construction process, though without robots, it is difficult to automate these processes fully. The age of the mobile robot is just arriving though replacing a man by a robot is not going to achieve the enormous savings of the production line. Perhaps with a full simulation model of the construction steps and a robot workforce, one should achieve better efficiency. In theory, one could approach the precision of the fixed-robot based production line, and one might get close to the maximum possible efficiency one might expect from building in-situ.

How do we achieve this with BIM? Probably all BIM systems are currently closed. The data needed is only partially available. So the vendors must make this possible through a mechanism that involves plugins and data access functions. With this, external vendors have an opportunity to add value to the model in real-time (synchronous).

As stated earlier, REfLEX and SONATA allowed multiple views (effectively a user defined plugin) that could access external system files and sensors, solve networks of objects, control external processes, generate live reports as the model monitored the real world, and produce reports and drawings that changed in real-time. It is not difficult. The issue is more about protecting

proprietary structures. It is likely that plugins will appear again in BIM systems, effectively making the systems more flexible, powerful and adaptable.

With these capabilities within the BIM model, many different aspects of Computer Integrated Manufacturing can be achieved. It will require specialist Apps or views, with access to all types of system information, including being able to change parameters and position data with the proprietary database. With this capability, we will have a better range of tools to manufacture, maintain and construct our buildings and infrastructure.

DIGITAL FABRICATION AND MODULAR CONSTRUCTION

One aspect of Computer Integrated Manufacturing that is used in the construction process is digital fabrication and modular construction (DFMC).

In moving from building to assembling in the construction industry, there is a fundamental shift from a business-operating model, where everything is unique and subject to extra cost, to a paradigm where much is repeatable and off the shelf, perhaps better described as sales orientated.

The AEC industry designs and constructs diverse structures with many components that are manufactured offsite and delivered complete. This includes windows, HVAC, electrics, and so on, but to achieve control, repeatability and savings, whole buildings must be automated. Complete coordination from an assembly and design standpoint must be achieved using the BIM model to allow these steps to be automated. BIM will control this automation, with the correct tools, via the timing and the precision and the interconnection between the systems. BIM also promotes the early specification of this work and the sharing of data between different manufacturers.

Some companies in the USA have been tackling this problem, albeit with limited tools, in particular, Katerra, FactoryOS, BokLok, Go Modular and others. It should be possible to achieve the offsite manufacture of complete building components using computationally assisted design and automated production and assembly, being limited only by the ability to transport those parts on site. One of the better-known examples is the Bird's Nest Stadium built for the Chinese Olympics. Indeed buildings are now being tackled that were too complex or risky to build onsite. Offsite digital fabrication has bought some of Gehry's buildings to life. Onsite construction led to inflated prices by contractors.

DFMC is the way forward.

---

3   "Computer Integrated Manufacturing and Engineering", Rembold, U., Naji, B.O., and Storr, A., Adcison-Wesley, 1993

Figure 210. *Waste in Manufacturing*

LEAN PRODUCTION

Over the last 50 years, the innovations of quality control systems, planning, scheduling, and manufacturing systems were integrated into the automobile manufacturing process. Of particular relevance to the Construction Industry are Project Management, systems integration, Just-in-Time, and Computer Integrated Manufacturing.

Looking at the Lean Production, JIT and Computer Integrated Manufacturing systems in more detail, we have powerful techniques that can be integrated into the construction process through the BIM model.

Lean Production deals with the elimination of waste through avoiding uneven work-loads and standardised work packages. In the words of Toyota, *"making only what is needed, only when it is needed, and only in the amount that is needed."*[4]

It has been suggested that Lean Production is the reason for the success of the Japanese car industry. There, Lean Production takes

different innovations in manufacturing, including Just-In-Time, Manufacture Resource Planning, Total Quality Management, Flexible Management Systems, and has redefined the production process.

Lean manufacturing involves the structuring, operating, controlling, managing and continuously improving production systems.[5] Restating the tenets of this philosophy in construction terms:

- Process stability- establish design and construction processes that combine man, machine, and materials to produce structures that meet or exceed all requirements of the various stakeholders.
- Standardized work-define best practice in each process and communicate this. This involves minimizing manpower and effort, improving quality of work and safety. Off-site assembly of components, where conditions can be controlled more carefully, is an important part of this. Workers make and own their own definition of best practices.

---

4   "The Toyota Way: 14 Management Principles from the World's Greatest Manufacturer", Liker, J.K., McGraw Hill, 2004

5   "Quantifying benefits of conversion to lean manufacturing with discrete event simulation: a case study", Detty, R.B., and Yinling, J.C. Int Journal Prod. Res., 2000, vol. 38, no 2.

- Level Production-balance and synchronization of all operations over time. This is achieved, in part, through the management of work packets and adapting the specialist skills on-site in a way that the process is done efficiently and smoothly. For example, having the electricians, the plumbers and HVAC services working in the most optimal sequence is critical to the optimal build time. Where each letter represents a process, the work sequence AAAABBBCC might be more efficiently carried out ABACABABC.[6] BIM should allow these work packets to be managed (and monitored).
- Visual control-status of design and construction so quality and timing problems are resolved. In the design process BIM achieves this by all stakeholders sharing and monitoring the model. Construction is currently visual. It could be scanned with and AI scanning system attached to the model, monitoring construction status. (The technology for this has just arrived).
- Quality-at-source-build rather inspect quality into the structure. Off-site manufacturing of large components, off-site quality control, inspection systems that provide immediate feedback such as AI based live camera systems and monitoring and control of problems that cause problems.

---

6   JIT Mixed-Model Sequencing Rules: Is There a Best One, McMullen P.R., American Journal of Operations Research, 2015

- Continuous improvement- empowers teams and employees to contribute their talents to continuously improve capabilities and processes.
- Just-In-Time procurement involves the delivery of the correct material at the right moment. This is essential on congested city centre building sites.

Discussing JIT further, in both manufacturing and construction, inefficiencies have been shown to occur because of stock shortages, incorrect orders quantities, improper storage, perishable materials delivered too early, multiple handling of materials, and out-of-sequence deliveries. It has been stated that the advantages of JIT in both manufacturing and construction include reduced inventories, smooth flow of materials improved quality and increased productivity.

Simulation, through the BIM model is a good way to implement lean manufacturing in the construction process. In addition the BIM model and process will assist with inventory management, transportation, and storage. Feed back strategies could be implemented on the BIM model to improve the processes on subsequent structure manufacture.

BIM can utilise the benefits of JIT. Visualisation helps clients and other stakeholders understand complex projects. Site layout storage and handling of materials in the BIM model solve

Figure 211. *Off-Site assembly of large components © Katerra 2019*

the issues of materials handling. Improved certainty and sure predictability also ensure uninterrupted workflow. Integration of JIT with the time model sequence assists with scheduling, storage, and reducing congestion onsite. Waste can be reduced; reduced waiting times, reduced transportation, and handling, processing waste, and inventory waste. Other advantages that BIM brings to JIT are better flexibility, better space usage and improved quality. Through a shared model, BIM also assists with information dispersal about deliveries, machinery maintenance, sequencing deliveries, accurate material quantities, and potentially physical limitations of delivery machines.

BIM has a very positive effect on procurement in terms of quantities, timing and source.

Winch argues that *"all products produced by discrete assembly industries, including construction, go through a distinctive life-cycle of: concept, design, planning and control, manufacture and assembly."*[7] In terms of construction and BIM this includes:

- *Concept* – a clear definition of the building in terms of client demand, achieved through sharing an early virtual prototype with all stakeholders.
- *Design* – the structure needs to be designed and detailed – the engineering is done through the shared BIM model and intelligent components, typically captured in engineering drawings
- *Planning and control* – construction processes need to be planned and then controlled according to the plan, typically 5D BIM
- *Manufacture* – discrete components and sub-assemblies must be transformed from raw materials into their final form, manufactured offsite using standardised modularised components.
- *Assembly* – the components must be assembled to create the finished structure.
- *Maintain and monitor* – support and maintain the building during its life, again through the BIM model.

Another aspect of Computer Integrated Manufacturing is what is known as DfMA: Design for Manufacturing and Assembly. This involves the combination of two processes, design for manufacture, which is the design for ease of manufacture of the parts making up the product or building, and design for assembly, or the design ease of assembly. These are two separate processes that must be applied simultaneously in order to achieve the most efficient and effective building.

In terms of BIM and construction, this is done through standardisation and modularisation of products that are used in the building. For instance, pre-cast concrete building components, pre-stressed concrete beams for buildings and railways, modular mechanical and electrical plant installations, and completed internal room and services 'pods'. BIM component concepts encourage modularity. Wide ranging components are being collected together into standardised libraries for use with BIM, encouraging the standardisation and modularisation to different degrees.

Unfortunately the standardised libraries are vendor specific, perhaps limiting their usefulness in some ways.

When standardised components and modules are used, they are assembled easily, and even better, pre-checked so that they can be assembled easily, leading to further wide ranging benefits in the design and construction of structures.

- Helps reducing costs
- Ensures improved quality
- Improved health and safety performance on site
- Higher quality construction
- Improved speed – up to 50% faster than traditional construction
- Earlier adoption of the latest innovations, fully tested and approved prior to commissioning
- Greater sustainability through advantages in thermal and environmental performance
- Lower maintenance and less waste generation
- Greater onsite recycling of materials leads to 'greener' construction outcome
- Greater efficiency in site logistics, with up to 90% fewer vehicle movements
- Manufacturing facilities adopting lean automation processes utilising the latest technologies
- Accurate quantities and accurate forecast of timings

Similarly suppliers, delivery times and delivery arrangements are all built into the BIM. In considering several parts that make up lean production, further insights into the possible improvements in construction design-build can be achieved.

---

7   "Models of Manufacturing and the construction process: the genesis of re-engineering construction", Winch, G.M., Building Research and Information, 31(2), 107-118

Future BIM

15

We have seen that Building Information Modelling has solved many problems in the Construction Industry. It has, however, introduced new complexities and challenges, and in some ways, has constrained original design methods. Its implementation is often complicated, requiring long lead times to understand and to gain expertise. At the same time, several technologies are available that might simplify the process.

The chapter on Retail Information Modelling outlines ways that these technologies are used to move to improved interfaces, interactions and achievement of purpose, albeit for different ends.

Here we expand on some of those ideas and others that the Author has experienced in the last few years. These are applied here directly to the construction industry. This chapter is initially concerned with interfaces, but then moves onto other technologies and ideas that can be used to help the complete design, construction and management process of BIM.

MAN-MACHINES INTERFACES

The majority of programs today are based around the idea of pull-down menus and icons. This idea is now forty years old, first seen in Bell Labs in the late 1970s. It worked well when the number of commands are limited, and the structure can be taken in at a glance. With too many menus, icons, sub-menus and sub-sub menus and hoards of icons, however, comprehension is delayed (perhaps permanently!).

The over engineered systems are too complicated (by definition?), certainly for the casual user. One needs an expert. One can have the designer telling the expert user how to construct the building, but that process is inclined to kill the design process.

The main contenders for changing the interface to the design process are the voice, language understanding, gestures, AI, AR or MR (Mixed Reality) and VR.

VIRTUAL REALITY

Virtual Reality offers a convincing simulated experience. It causes the brain to believe it is in an artificial world. It is currently used mostly for games, it has never been taken up to initial expectations. It is used to in BIM usually to give the client a realistic, in-there view of the building. Figure 165 shows some of the original VR work in SONATA, and Figures 185 and 197 show live images from the VR program, where the customer can interact with the environment. Both sets of work were work done by the Author in 1995/1996 and in 2019/2020.

There are various problems with VR that preclude its prolonged use in design, the most common being an over intense experience. One feels enveloped in this world with all other visuals excluded,

Figure 212. *Microsoft Hololens 2 AR worn by co-RIM developer Mark Edwards*

and this can lead to tiredness, nausea and dizziness. It is all too much over any length of time. Some suffer from a form of seasickness. That notwithstanding, it has been used in many applications successfully. The Author had the best quality VR glasses (for 2018) on his desk over eighteen months for development and, in the end, avoided the use of them as much as possible.

In the retail industry, it is used to generate research information about customers' shopping habits. The customers are sent into a store to shop in VR. They can select and examine virtual products, answer questions, eye viewing direction is tracked, and movements around the store are known, all essential to understanding what drives the customer. (Figures 193, 198, and 199).

But as a design tool where immersion is required for days at a time, it is too intense. This has not stopped various products being built to achieve just this. Unity and Autodesk (Revit) have combined forces to produce Reflect[1] that places a live BIM model into a 3D games environment, including VR. This is indeed interesting but again relies on Revit to do the changing of the model.

In terms of interface design, AR offers better, albeit less immersive, capabilities.

AUGMENTED AND MIXED REALITY

AR presents virtual objects that merge with the real world. These objects can attach to real-world items and can look indistinguishable from them. Mixed Reality is Augmented Reality where physical and virtual objects interact. It is a hybrid reality. For simplicity, Augmented and Mixed Reality are treated as one, Augmented Reality (AR).

AR and VR are very different tools; AR provides greater flexibility in its use, it is less intense in its presentation and hence allows the user to use it for more extended periods . It also allows the user to see the stabilising world and allows the user to interact with the real world in a way the VR cannot. An integral part of the AR tool is the detection of real-world objects and surfaces, and in some systems, includes visual positioning. Visual positioning involves knowing the position of the camera at every moment using "hook points" into the real world. The smartphone camera uses the accelerometers, the compass and the visual data to determine how it has moved relatively. How can this be used in BIM?

_____

1   https://blogs.unity3d.com/2019/06/05/unity-reflect-bim-to-real-time-3d-in-one-click-for-better-design-decisions/

Figure 213. _UnderstandingBIM app available in your app store. Live AR appears when the camera is pointed at the pages of this book. There is an index in the App to show which pages work._

The earlier chapter Retail Information Modelling details the use of AR in retail applications. In these applications, in AR, we determine and check the positions of products, we layout the walls, detect windows, doors and other fixtures and create a model of the store in detail. Another application uses Hololens 2 [the Author's arrived today] and voice control to build a complex planogram and the movie "Minority Report" illustrates this.

One can easily envisage using an AR system such as the Microsoft Hololens 2 System and Google voice systems to construct a

Figure 214. _Both are live and interactive, which is virtual? Unity London Office © Unity 2019_

building. Gestures, voice commands, and perhaps a screen will allow the user to use the system intuitively — the voice commands, discussed in the next section.

UnderstandingBIM is an app available in the various app stores, built explicitly for this book. AR is used to bring the book to life with models and animations referenced in the book.

The possibilities for AR in BIM and the CI are vast; the list touches almost every aspect of construction and is enhanced by BIM. It will include, but not exclusively:

- Conceptualising and designing the BIM
- Reviewing costs
- Sharing and presenting concepts over a boardroom table
- Sharing concepts on site
- Assembling model concurrently
- Putting live hooks and tools into the model
- Sharing the spaces
- Monitoring sensors in a finished building
- Monitoring maintenance
- Reviewing Assets
- Helping with directions
- Safety
- Examining HVAC in ceiling spaces
- Checking flow rates
- Monitoring crowds
- Examining area usage with heat maps

and so on.

If you add AI into this equation, you get a powerful intuitive simple system. The implementation of the RIM systems discussed in Chapter 12, outlines some of these possibilities.

The most significant difference would be in replacing the layers of pull-down menus and icons with a gesture, voice system. This would enable the designer and contractor to share the space in either the virtual or real building, move and replace objects, connect objects, change detail with gestures and voice. Chapter 12 and the AI section later in this chapter expands on these possibilities; not just possibilities but actualities in the case of RIM.

GAMING ENGINES

Gaming engines, in particular, the Unity and Unreal systems, provide an excellent development environment for design and construction. Initially designed for games, there has been a push to see them used in the CI. Fully 3D, they provide the developer with a powerful set of tools across a range of platforms, with wonderful visualization and physical interaction between the

objects making up the scenes. They are optimized to provide the very best graphics from the modern graphics processors (GPU) and provide exceptional tools into VR and AR environments. In a joint program with AutoDesk, Unity provides direct access into BIM models and multiple AutoDesk formats. The images generated are difficult to separate from the real-world equivalent. Figures 217 through 219 illustrate this very well. It is difficult if not impossible to tell which is the photo and which is the live image generated in Unity. In addition to providing different types of interaction with the users, these environments allow apps and programs to be built for a large range of systems. It is easy to generate a system for a PC, as an iPhone and as an Xbox.

The gaming engines provide multi-platform support with a host of lighting, reflections, textures and other tools. This simplifies the task of the developer, extends the functionality across multiple platforms and allows multiple users to share the same environment.

VOICE AND LANGUAGE

Speech is a simple, natural and powerful way of communication. Understanding the spoken word in terms of word matching and then understanding intent has steadily advanced in the last decades. We come across these daily. The good ones are great, you barely notice, the bad ones are excruciating. You can get insurance, order pizzas and many other things.

Two processes must be undertaken to understand and act on the spoken word. The first process converts the speech into written words and the second "understands" the words and derives an intent from the spoken sentence. The former deals with the multitude of languages (Google has 120 currently), and the second the derivation of intent and entities. The intent is the verb, what must be done, and the entity is what it is being operated upon.

The Author has been using Dialogflow, Google's voice for understanding and interpretation to design retail interiors and laying out products. The accuracy of understanding can be seen using the microphone button on the Google search. Google has announced that the language interpreter can identify and follow multiple voices speaking at the same time and determine the mood or character of the speaker. Understand means convert the spoken word into its written equivalent.

Both Google and IBM provide language analysis (DialogFlow and Watson) that enable this "understanding" and, as such, can provide a general non-boring interaction. Neural Networks are used in these processes.

The system needs to understand the myriad ways you can ask for something.

"I want to place a brick wall 5 metres long and 200 cms high."
"Build a standard partition so it is 5m long."
"I need a wall. It should be 5m long. "

Here the intent is build, and the entity, in this case, is wall or partition.

"Show me the windows I used on Canary Wharf project."
"I want to go to the beers," followed by *"I want a cold one."* (indirect reference)

"I love you." (irrelevant chit chat)

What is the cost of this fixture? (live data and identifying objects in the scene)

"Where are the loos?" (directions and slang)
and
"What windows did I use on the Canary Wharf Project?" (historical data)

and so on. There are many many different possibilities for these questions, but the intent (verb) and entity (subject) structure together with context capability mean "understanding" is possible.

In the retail application, there are potentially upwards of 50,000. DialogFlow has no problems in differentiating between them all, coming back with the appropriate barcode. Every country has different product ranges and barcodes.

Google claim to recognize the mood of the speaker and can also identify and interpret a single voice in a crowded room.

The BIM equivalent might be BIM objects, with supplier modifiers, or areas in the buildings. Place something is the intent. So when the user asks in some way for a wall to be placed, the system performs appropriately. There are issues of context, who the user is, the area being worked on, the current preferred supplier, the local codes, the use of the building, the weather (perhaps), the climate, the orientation of the wall, is it internal or external and the last walls used. The AI section will discuss understanding how you have used these before and how to allow for these contexts.

The science of language understanding and recognition is improving and, as the systems are used more widely, the neural-network-based recognition algorithms will improve. The aim in the retail information and in BIM should be to explain to the system what you want and have the system build the model. It will automatically join and set the BIM parameters as if an expert had done it. The user could then use as little or as much of this capability as desired. There would be a significant AI behind this capability.

AI AND CLOUD-BASED GPUS

There has been a significant increase in the use of Artificial Intelligence (AI) across all aspects of our lives in recent years. We use almost self-driving cars, we chat to chat-bots on the phone without even realising it (the good ones), and AI has beaten humans at every single game devised by man. This revolution has been a long time in coming in that AI has been the subject of intense research for perhaps many decades. The "Turing test" determines if a particular AI is capable of thinking like a human. The Turing test was passed by a robot named Eugene in 2014.

AI essentially works around the neural nets. Neural nets are found in the brain in cortical columns. There are some trillions of cortical columns in the brain and these are grouped together to deal with the different senses and reasoning. They are all similar to each other.

Neural Nets are used to recognize patterns, images, writing and voice. For example, they have been used to find patterns of cancer in a population. The best solution to date has come from TensorFlow.

TensorFlow is a library or system used to train neural networks. In technical terms it is a "gradient-based optimization and training of deep network". It was created by Google and is now open source so anyone can look at the code, how it works and make contributions. One of the main breakthroughs of TensorFlow was that it allowed neural networks to be trained on Graphics Processing Units found in every modern computer. These are normally used to make high quality graphics for animated games. Very powerful versions of these are now found in the cloud, enabling high quality graphics games to be played on any machine anywhere. Tensorflow utilizes these processors across multiple machines by splitting the network and data.

This enables neural networks to be trained in days instead of weeks.

How does it work? It does this by creating objects out of the data (explicitly knowing everything e.g. shape, type etc. of all the inputs) and also by using a "data flow" graph which allows the network to understand how all the functions will act on the data allowing it to organise itself across multiple GPUs/computers accordingly. Note the word objects. BIM objects, their data, their position orientation and relationships to each other are an ideal form of data for TensorFlow.

Artificial Intelligence is defined as the system enabling machines to perform tasks requiring human intelligence. It works but utilising large amounts of data to "teach" a machine how to perform. It is diverse and has many applications from understanding speech, to driving cars, diagnosing medical conditions, recognising images and handwriting, and so on. AI is ubiquitous, being found in most sophisticated technologies in one form or another.

Sometimes it does this utilising a neural network, a technique that emulates the brain's cortical columns. Possibly of interest is that the Author spent several years using a variant of an object-based design system to assemble cortical columns and pass information (synaptic impulses) down the columns. Building cortical columns out of the very BIM objects that had been developed during the last twenty years. The initial implementation was a single neuron per object. Because of speed, this was rapidly replaced by having a simple neural net per object, with information being passed between the objects in very much the same way as described for the networks early in the book. The issue was that it was too slow to do anything or to scale.

This hardware allows us to create a BIM framework that is "optimised" to run on GPUs across multiple machines in the cloud. These are normally used to generate the high quality 3D CG images we see in movies, advertisements and in games on your computer. These cloud based GPUs are incredibly powerful and prolific and take the computational power to the next level. So to recap, the technique we use to learn and solve the problems of building design (and many other things) is called TensorFlow. This has been optimized to run on powerful cloud based GPUs enabling us to solve very complex design and optimization problems.

So imagine if we "fed" the TensorFlow model with perhaps 50 or more modern hospitals. We then lay out a foot-print and some other site restrictions (climate, height materials) and create a new hospital automatically, in full detail, using the TensorFlow techniques. The system has learned what a hospital should be like and in theory, it can be designed down to the last nut and bolt, the last mirror in each bathroom and the size of the duct work and steel. It could produce a BIM model. In RIM the Author has built simple layouts and had some success.

Similarly let us try training specific designs of a well known architect. Specify some constraints and, hey bingo, we have a building in the style of your famous architect.

After feeding buildings designed by Gaudi and Norman Foster through this AI system it would be interesting to see what sort of designs the system would instigate.

Out the other end comes buildings we have never seen before. Is this feasible? Absolutely yes. And as the machines become more powerful the level of detail of design will increase. PennyLane is a different system similar to TensorFlow but has been designed to be fault tolerant and to run on a Quantum computer. The technical term is "variational (quantum) circuits"[2]. The term QNN (Quantum neural networks) has been coined to designate this type of operation.

The musical equivalent is to feed in the works of Bach into the networks and get the machine to compose a new work. This has been done, with moderate results. In the Author's opinion, these works sound like Bach but do not have the brilliance of his great works. In a way, this problem is more difficult than a static building because of the interrelated temporal effect between consecutive bars of the music, let alone the emotional content.

In buildings this is not the case though still difficult. A way around this is to generate a huge number of buildings by Gaudi-Foster-AI or pieces by Bach-AI and choose which ones you want. One might make a movie of evolving parameters. The problem is then compute power.

Another possibility is what is called adversarial generative design. Imagine generating a host of buildings similar to the one required and then optimizing the characteristics in terms of cost, materials, construction time or capacity.

Once the structure is created, one could use generative design for the interior layout. At every stage of the designing process have the ability to allow AI to give you inspiration as to what is possible within your constraints (budget, space etc.). This is done in mechanical engineering to optimize the weight of materials used in mechanical parts leading to interesting and complex shapes that need to printed with a 3D printer.

---

2   "Progress towards practical quantum variational algorithms", Wecker, D., Hastings, M. B., and Troyer, M., Phys. Rev. A 92, 042303, 2 October 2015

AI can allow one to generate a 3D model or layout from a simple 2D drawing. At the simplest level this involves drawing a sketch of say a chair and the system will actually create a chair. Similarly a quick sketch outline of say a bathroom can be recognized as a full bathroom and the appropriate design dropped into place.

AI also has the ability to redesign everything to fit together after changing a component in the design. For example, if we move an exterior wall of a building then all of the model around that will be updated and adjusted accordingly.

AI can be used to automatically generate a build sequence for a building mode. It would do so "knowing" delivery times, how long different processes take, the order of the construction sequence, and limitations with the number of people working. It can produce the GANTT chart and a construction video sequence, for any set of constraints. AI has the ability to take into account expected weather and statistically vary the construction timing based on different weather events. Print out the report of the time-line and give to the project manager. Maintaining the exact finish time (a bit like ETA with a SatNav) is easily done as the current status is maintained. Any variances in construction time will be learned for the next build. It will understand all the issues which can cause an increase in time before the project has started; AI has learnt these from previous development projects.

The AI can be taught to understand about materials, properties, where to get them from and how much they will cost. Get them ordered to fit in with the time-line mentioned above.

AI should be capable of generating a very accurate schedule of costs, taking into account all relevant criteria. Using the construction sequence criteria from above plus parts, plus perhaps even predicted price changes, a very accurate estimate will be achieved. The AI will learn by reviewing the existing building, costs and construction times.

A carbon emissions AI could be run to determine, very accurately, the carbon cost of a building components, the cost in construction and the carbon cost in running the building (see Cloud Based Physics Section).

It could scrape the Internet knowing the materials and components making up the building, together with transport emissions and so on to come up with an accurate picture of the carbon cost.

Predictive behaviour of building components can also be undertaken by the AI. Rather just knowing about previous building maintenance issues, the AI would do much better with sensor input into the model. This is discussed in a later section in this chapter. Monitoring behaviour of the real world equivalent through components in the model was something we did in the 1980s. See Mott McDonald in the Engineering Chapter 10.

This list barely scratches the surface of what is possible. The potential for AI across the whole design process is huge particularly when linked with the compactness of the BIM model. The style of the BIM model and the data requirements for AI complement each other.

CLOUD-BASED PHYSICS ENGINES

A physics engine is a part of a GPU (usually) that computes calculations to do with the physical world in hardware. This has been developed for the gaming industry to allow realistic interactions between bodies. Distributed computing allows for a more advanced physic engine (as seen in cloud-native gaming) which can predict stresses more accurately as well as adapt to designs more quickly. It is not clear which is the better way to solve the problems.

There are two possibilities for predicting the thermal performance of a building using the new technologies:

- use a learned performance from known building models as discussed in a previous section.
- use the physics engines to model the flow of heat across the different BIM components.

Physics Engines might also be used to understand the impact of the heating, solar radiation, cooling system, natural habitat, light and related physical interactions.

In Health and Safety the physics engine could be used to predict the effects of a fire at every single location in the building and use AI agents to predict how people would escape the fire. Every possible location of a fire could be tested and a safety report could be produced with the predicted number of fatalities. Furthermore, use the AI agents to examine the flow through the building to make sure that no crowds form. Crowd simulators have existed for many years.

THE OPEN BIM ENGINE

"Open BIM is a universal approach to the collaborative design, realization, and operation of buildings based on open standards and workflows." Open BIM seeks to allow all vendors and users to share information via a common data structure.

Using BIM systems from different vendors is problematic as communication direct between the models is non-existent. Various techniques have been introduced to assist in this process. These have concentrated on what is outside the system, i.e. setting up external files to pass information between the system. One of these techniques are IFC files and are discussed in Chapter 4 above. Unfortunately IFC files do not capture all the information such as parameters, connectivity, and so on and they can be out of date as soon as they are generated. Additionally there is nothing automatic about the transfer of information between the various systems, so sharing when different stakeholders have systems from different vendors is problematic.

A different approach, certainly if you have the advantage of building a new BIM system, is to make the files open to all. The capability of writing into the BIM data structures from almost anyone brings a range of new problems, but it is possible to manage this by providing access to "approved systems", a bit like the Apple app library. Alternatively, a library of functions might be provided to allow full controled access to the database.

Figure 216 shows how data might be structured to achieve this visibility, shared functionality and access. (as opposed to Interoperability). It brings all the capabilities needed, possibly delivered by different vendors and stores the information in the instance database.

The open data structure might be through JSON (Java Script Object Notation) that is widely used across different systems. Without going into it, this is a text-based, data structure which is easily loaded directly into a modern computer program without change. The structures are obvious and are easily assembled.

Open BIM might contain some binary data or instance records. If this were made generally available to all Open BIM users, it would preserve the integrity; it would promote efficiency and allow wrappers around the wrappers to add extra functionality retaining the openness of the file system itself. For instance, if a particular vendor could access the data structure efficiently, then perhaps it could be inserted between the Open BIM instance record calls and the user application calls. (Note: One of the problems of Revit is that changes made to the library are lost on the annual upgrade.)

Another aspect of Figure 216 is that the components themselves are not defined as part of the Open Data Structure: components or BIM objects are defined via plugins. In technical terms, this means that the BIM objects are compiled and linked into the system at run time. Different types of views can also be added to this container. Individual vendors could write their own plugin for

any particular component, potentially allowing standard office software to access key attributes of the BIM.

What this means for users seeking access to data, say the steel fabricator, can write pieces of code for each component in the system, effectively allowing the fabricator's software to read directly from the BIM, add his own piece of code for each component. These become self fabricating components. The system provider (for example Autodesk or Graphisoft) would provide a library of a series of function calls that allows him to "view" each of the components within an area of the building, perhaps by layers and volume. Usually these layer/volumes are named, where each component executed using the fabricators code, or "view," will do what is needed. All data available to the component would be available to the fabricator without copying, so no data duplication would be required.

Advantages for the fabricators, apart from knowing everything is up-to-date, are that they can perform calculations and look at local information in the BIM in their own environments.

This could work for any number of users and is not restricted to fabricators. For system developers, the changes are minimal, but fabricator/users would need to have access to all of the functions available within a parametric component. This mechanism becomes crucial for Level 3 BIM, where there is a "Single Source of Truth" from which everything should be possible. There can be no copying data or structures into a format based on Standards if this is to be achieved.

With the Open BIM structure comes the opportunity to change the way the database functions. True cloud distribution can be achieved using systems such as MongoDB. The database can be scaled without performance loss and secured in a way that an SQL or file based system cannot. MongoDB scales horizontally and uses a system known as "sharding".

PLUGINS

Plugins will become very important. Plugins allow vendors to add value, different disciplines to achieve their ends and be self-contained and proprietary. There are no longer Lowest Common Denominator issues. The Building Information Model becomes a database of elements, each of which has the hooks to deal with itself.

The same idea could be used for QS and estimators' man-hour costs. Each firm of estimators would have a view associated with components making up the BIM. For instance, in estimating the man-hours costs to install a particular duct network in a structure,

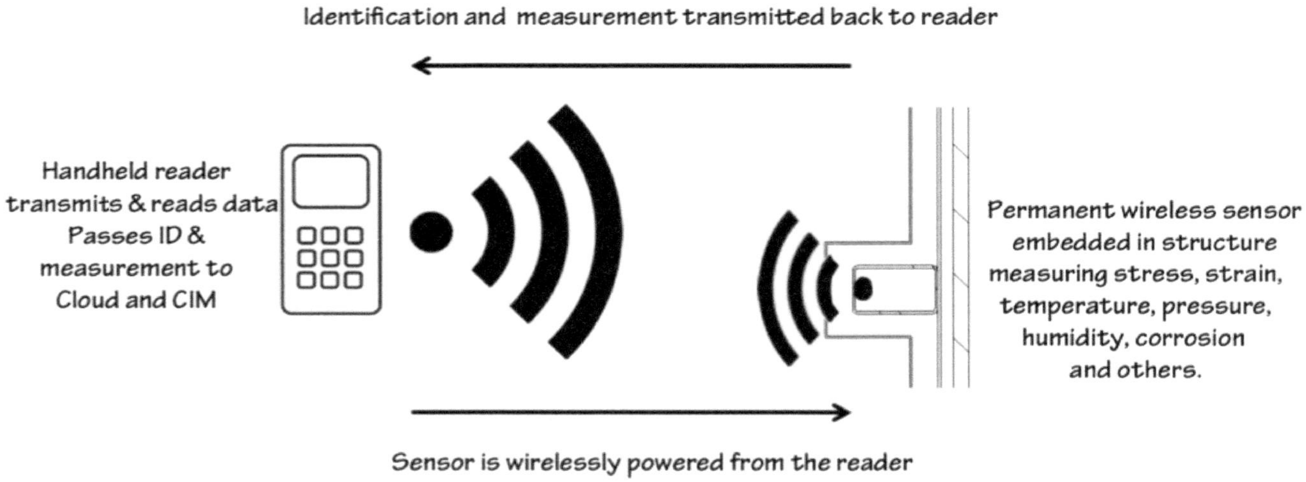

Identification and measurement transmitted back to reader

Handheld reader transmits & reads data Passes ID & measurement to Cloud and CIM

Permanent wireless sensor embedded in structure measuring stress, strain, temperature, pressure, humidity, corrosion and others.

Sensor is wirelessly powered from the reader

Figure 215. Technology of permanent built-in wireless sensor, to measure physical variables

the "estimators" view would look at the size of the duct, the material, any special installation requirements, and with a bit of cleverness, how difficult it is to install. It becomes a part of the duct. This information would all be written back into the estimator's package, perhaps MS Excel. The information comes from the instance information and surrounding components.

Another use for this Plug-in is that the "fabricators" would use their own "Physics" views. What this means, for example, is that HVAC contractors bidding for work can use their own design rules within the building model. They would be able to solve complex duct networks in their own ways to win work by designing more efficient networks based on their direct experience. It would allow the user to solve networks of components, including stress, strain, flow, heat, etc. in their own particular way, for which they know the costs. Incidentally "physics views" were included in SONATA and REfLEX.

Some effort goes into generating all of these specialist views, but once done for a component set, data is fully inter-operable between disciplines. There is plenty of scope for different firms to differentiate themselves from their competitors to win work through better estimating, and it would give everyone more confidence, refining their view of the various aspects of the building costs.

OPERATING BIM OBJECTS AND BUILDINGS

A system to manage and operate a building is achievable with a combination of AI and dynamic building model. Borrowing some

of the ideas from the RIM and City Information Model from earlier chapters, it is possible to envisage a system where buildings are run and managed optimally, with minimum human interaction. If we consider first operating and maintaining a building from the extended BIM concept.

The plan is to run the building in the cloud alongside your actual building. As the real building changes (lights go on/off, heating on/off etc.), the virtual building also changes. Sensors are used around the building to understand the effects of heating, windows being open and similar, and to continually update the AI with this information. As in the RIM store management idea, it is proposed to have an AI running the maintenance, building services and security.

Reintroducing this functionality would bring a number of benefits to the functionality of the BIM model. This functionality ties the BIM to the operating real world and could extend the use of the BIM into many dynamic situations.

By extending the functionality, this would allow elements to write to external devices such as lights, locks, security, movement, airflow, temperatures, lifts, or in fact, anything that needs to be switched on-off, or operated in a structure.

Extending the idea still further, one could have an installed state of health view of the building based on the sensors (last read values) in the building. Stress, strain, humidity and noxious gases could all be measured with readers that feed directly into the BIM. Thermal performance becomes a critical part of the cost of

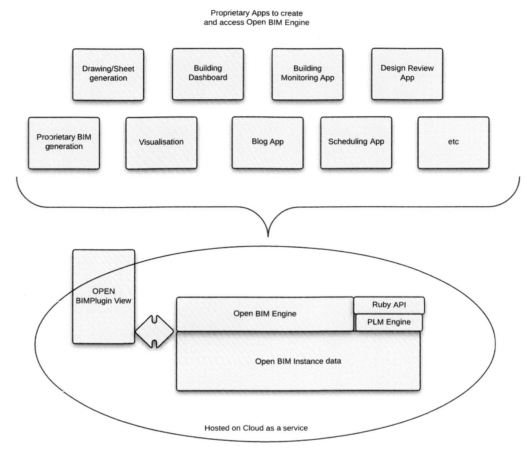

Proprietary Apps to create
and access Open BIM Engine

Figure 216. *File Format for the New Open system definition for BIM*

running the building, not only in terms of the actual heating costs, but also in terms of the environmental costs. Various sensors have been developed with the specific purpose of monitoring building and structural performance. Continuous monitoring of stress in structural members, using fibre optics together with other types of sensors, is already in use on major bridges.

Having a sensor permanently in structure has been limited to accessible parts of the building. The Author has patented a new technology that allows permanent sensors to be embedded in structures. These are very small, very-low-cost, battery-less sensors requiring no maintenance. They are based around a simple RFID chip and aerial, transmitting at different frequencies depending upon the property being measured, using specifically chosen MEMS[3] sensors. Pressure, temperature, stress and

corrosion can be measured using this passive technology. Sensors are very cheap, typically less than £1, permanent and can be embedded in walls; you could come back in 30 years to measure some property. RFIDs tell the reader what is being read, and the frequency determines the property. Systems are currently being trialled for measuring the pressure and temperature in truck tyres, but other applications within building structures and infrastructure include

- HVAC - flow, temperature, pressure and humidity with ID to be placed inside HVAC ductwork to measure these parameters at any time during the life of the building.
- Stress/Strain sensors to determine long term changes in building structure, even embedded in concrete.
- Movement sensors to detect small changes in building.
- Corrosion measurement in voids, reinforced concrete and other inaccessible voids, etc.
- History or previous readings.
- ID of each device.

---

3   MEMS is a Micro Electro Mechanical System, a tiny electronic component potentially with moving parts that is part of the electronic circuit.

Figure 217. *Computer Generated Images © 345 Holdings 2019*

Figure 218. *Computer Generated Images © 345 Holdings 2019*

When activated by a handheld reader, parameters are read and transmitted into the cloud as an ID and sensor values. This information can be displayed in the model as data, graphics, warnings, or even have the specific sensor object send an email.

Preventative maintenance, often based on the number of operating hours, can be monitored by the BIM objects. They may refer to connected cabling or switch positions to understand when the device is running.

Time-based maintenance is much simpler with a view that executes at regular intervals. For instance, an escalator may need to have various tasks performed (say) every 12 months, greasing, cleaning, checks, and so on. The virtual representation of the escalator will need to be executed once daily to ensure that the maintenance date is maintained. In early BIM systems a keyword

associated with the component that caused the system to issue a notice when doing maintenance. The BIM object itself issued the appropriate warnings to the concerned persons.

FINITE OBJECT MODELLING

In the 1980s, the Author tried some experiments with objects from the BIM model, which involved making a mesh of objects and solving stress and displacement. This section reintroduces this idea known as FOM.

Many objects are joined together in a BIM with some meshed and others fixed or pinned. Each object passes displacement and stress information to its neighbours. Some objects might represent a pin, others a rigid connection that passes moment, displacement and shear force. Some objects might be part of a mesh with connected nodes, while others might represent beams and columns. Each object has a "physics view" that defines how it

Figure 219. *Computer Generated Images © 345 Holdings 2019*

behaves given the forces applied to it. Each time the physics view is executed, the forces and displacements are calculated given the values of the surrounding connected objects. This is similar to the duct network seen earlier in the book. A solution to the distribution of displacements and forces can be found repeatedly executing all the interconnected objects until the forces and displacements stop changing. Behaviour may be linear or not; it may be driven by tables, external rules or criteria as the applied forces change, the strength of the object can change, changing its stiffness and weight. These are adapted into the framework of the solution. The objects themselves are neither limited in shape nor size and may connect to adjacent or non-adjacent elements in a plethora of ways. Dynamic behaviour and flow might also be tackled this way. The Author has referred to this as Finite Object Models (FOM); a Finite Element Mesh with Objects.

SUSTAINABLE BIM

Climate Change looms and brings potentially unmitigated disaster. We in the construction industry must change our ways in design and construction to eliminate waste, reduce GHG emissions, reduce pollution, reduce energy usage, and increase the usable life of buildings. We in the Construction Industry are responsible, in no small part, for climate change.

The UN puts the position of the building sector rather succinctly: *"Globally, buildings are responsible for 40% of annual energy consumption and up to 30% of all energy-related greenhouse gas (GHG) emissions. The building sector has also been shown to provide the greatest potential for delivering significant cuts in emissions at low or no-cost with net savings to economies. Collectively the building sector is responsible for one-third of humanity's resource consumption, including 12% of all fresh-water use, and produces up to 40% of our solid waste. The sector also employs, on average, more than 10% of our workforce. With urbanisation increasing rapidly in the world's most populous countries, building sustainability is essential to achieving sustainable development."*[4]

These figures are significant in their contribution to the impending disaster and so must be addressed. Other factors that must be addressed include responsible material sourcing, banking and recycling of materials, improving air quality, facilitating a shift to sustainable transport, and supporting green infrastructure.

The Paris Agreement pledges by governments in 2015 are consistent with warming of 3C by the end of the century. 3C is probably a catastrophe. To achieve a more reasonable 1.5C rise would require emissions to fall by about 50%[5] between now and 2030, and reach zero by 2050. See Figure 220. At three degrees rise, the polar regions are severely impacted, coral is dead, some ecosystems are threatened, and coastal flooding is significant, droughts and resulting fires in Australia are likely to be intensified by climate change.

Greenhouse Gas (GHG) emissions caused by the Construction Industry are significant in the global scheme of emissions. By far, the most significant contributor to the total for the CI is the operation of the building emissions at 72-83% depending on whose figures you use. This is perhaps as much as 40% of the total world-wide GHG emissions across all sectors. An additional issue is that the number of buildings in the world is going to double by 2060, effectively an entire New York city every month for 40 years.

There have been improvements in the energy efficiency of the construction industry, but these have not been nearly enough to offset emissions from new construction. GHG emissions are continuing to rise from the CI at the rate of 1% per year.

To limit the rise in global average temperature to 1.5-2.0C, all new construction must use no CO2-emitting fossil fuel energy to operate (or GHG collected), and new buildings must be highly energy efficient.

Another factor that affects the total GHG emissions for the CI is embodied carbon. Embodied carbon emissions are those generated when constructing the building and its constituents. The production of iron, steel and concrete account for about 9% of the annual GHG emissions. Cement is particularly bad, not so much in quantities of CO2 per tonne (0.107kg as opposed to steel 1.46kg) but instead because of the much greater amount of cement produced.

So buildings must be durable as well as efficient. The most durable buildings will last hundreds of years, removing this embodied GHG cost, or at least spreading the GHG cost over those years.

Extending BIM and applying across the Construction Industry (CI) has the potential to significantly reduce GHG emissions globally

---

4   "Global Status Report", UN Environment, 2017 https://www.worldgbc.org/sites/default/files/UNEP%20188_GABC_en%20%28web%29.pdf

5   Rather than use MtCO2e as a unit to measure carbon dioxide measurement, this chapter will deal in percentages.

by perhaps as much as 10% or possibly 15%. The UK Government, in its early enthusiasm for BIM, suggested that emissions could be reduced by one half. This equates to 20% of ALL emissions. The Author is proposing an approach similar to the car industry where all buildings must meet rigorous standards to solve this problem. The Boston Consulting Group has projected savings of over one trillion US dollars a year from using BIM in the CI[6]; those savings would contribute significantly to reducing GHG emissions.

The Author has formed a small group to enable several things to happen:

- To specify and assist in the provision of a Carbon View for BIM Objects to evaluate the embodied carbon cost of the building. See Below.
- To specify and assist in the provision of energy performance standards for buildings, including monitoring and maintenance.
- Approach the governing authorities of those countries where BIM is used to enforce the presence of Carbon View in BIM objects.

_____

6 "The Transformative Power of Building Information Modeling", Gerbert, Catangnino Rothballer, Renz and Filitz, Boston Consulting Group ,2016

- Approach the OECD, and IEA and others to have the built-in standards enforced.
- Approach major BIM systems providers to enforce the introduction of Carbon views.

CARBON VIEW

We have seen that Objects or components used in a BIM model have different views associated with their structure. These views represent various aspects of the objects that are presented depending on how they are viewed. To remind the reader, a physics view was added to objects to solve the physics of fluid flow, temperature, and pressure etc. Here, this view might be used by the system to solve the problems of thermal flow in the component parts of the building.

In the Carbon View, information is assembled to determine the embodied carbon cost. This is a complex calculation and is different for every component making up the building. For instance, the BIM objects representing concrete will have a CO_2 GHG associated with the making of the cement, the mixing and transport of the concrete, and so on. The quantities, type of concrete, the hoisting height and so on, is known for each object, similarly, with the steel in the building. A telephone, a urinal, or a

Rising temperatures, rising risks

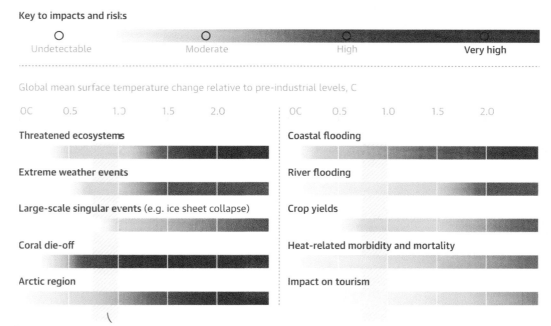

Figure 220. *Rising Temperatures Rising Risks courtesy Guardian Graphic IppC Special Report on Global Warming of 1.5C*

hot water system all have very different carbon costs depending upon their manufacture, the origin of the materials etc.

Some of these costs might be recurring. (It is more convenient to put them here even though they are not true embodied-carbon costs). For instance, maintenance, including transport, parts, material use, etc. might be included.

The view or plugin aims to minimize the energy cost of buildings over the long term concerning heating, lighting and air conditioning principally, though additional items such as hot water, electronics and refrigeration might also be measured as part of the process. A the same time, it must predict or learn to predict the performance of buildings to assist the designers in making the process more efficient. By necessity, this is a two-step process.

A sensor view or plugin attached to a BIM object could read smart meters and built in sensors to measure the performance of the building.

Predicting the performance of a building involves using known software and building an AI system to take basic parameters

from the sensors and meters as well as the climate, the building parameters, and other energy information about the building.

LEVELIZED COST OF CARBON

We need a tool to compare the different sources and forms of GHG emission over different timespans. The idea of a levelized cost of energy (LCOE) comes from the power industry, and we adapt this to a new tool, levelized carbon cost of construction (LCCC). In the power industry, this tool allows diverse costs and forms of energy generation, over varying lifespans, to be compared and bought back to a single rating. The levelized cost method would enable one to compare nuclear power plants with a few solar panels on the roof, a wave turbine with the coal-fired power plant down the road. (The Author spent several years designing and building wave turbines continually comparing costs and energy production).

There are many different environment costs; embedded carbon in the materials making up buildings, the running costs, transportation costs, energy costs in operating the building, maintenance costs (environmental) and so on, all of which need to be summed.

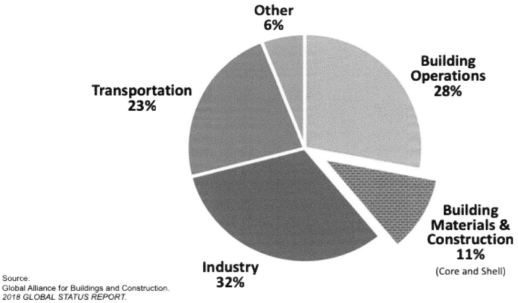

Global CO$_2$ Emissions by Sector

Other
6%

Transportation
23%

Building Operations
28%

Building Materials & Construction
11%
(Core and Shell)

Industry
32%

Source:
Global Alliance for Buildings and Construction.
2018 GLOBAL STATUS REPORT.

Figure 221. *CO2 Emissions by sector Courtesy Architecture 2030, 2019* https://architecture2030.org/buildings_problem_why//

In an extended BIM environment, with extended views or plugins, these calculations of carbon cost could be automated within the components themselves. The total carbon cost/year is equal to the sum of them within the BIM, as follows;

LCC(tonnes/year) = (Cembed + Cassem + NPV of Cuse + NPV of Cdemol- Cvirgin +NPV of Cmaint + NPV of Cwaste + NPV of Cresources)/LifeOfBuilding(years)
- LCC levelized cost of carbon
- Cembed embedded carbon cost of all parts + delivery + etc.
- Cassem carbon cost of assembly
- Cuse carbon cost of use (this value depends upon how energy is generated over the future years and transport etc.)
- Cdemol carbon cost of demolishing
- Cresources
- Cvirgin MINUS the cost of leaving the site untouched
- Cmaint maintance, refurbishment etc. carbon costs
- Cwaste waste disposal, storing.

The NPV calculation in the LCOE energy calculation is determined by the interest rate each year. In this LCCC calculation, it could be the cost to the environment in each year.

The Author is suggesting that all buildings must have this and if you are using a BIM model, then each BIM object or component has a plugin or separate view. This plugin will provide the ability to generate ongoing carbon information about the object. This calculation on all objects will lead to a more straightforward estimate of building carbon performance. Making such a number

Annual Global Building Sector CO$_2$ Emission

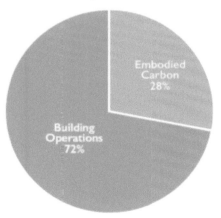

Source: © 2018 2030, Inc / Architecture 2030. All Rights Reserved. Data Sources: UN Environment Global Status Report 2017, EIA International Energy Outlook 2017

Figure 222. *Construction Industry CO2 Emissions. Courtesy Architecture 2030, 2019* https://architecture2030.org/ buildings_problem_why//

obligatory via legislation will ensure that it happens. It is essential to have a means of comparison between diverse structures and infrastructure.

Given the discounted carbon cost of the building we have a value that can be used to rate different designs in a comprehensive way. These figures provide an annual cost, carbon and monetary, and can be divided by each other to give the carbon cost per dollar, allowing designers to evaluate the real cost of carbon reduction.

It may be worth including the effects of reflected and absorbed heat, as a sufficient number of reflective roofs makes a difference, and so will be added as CO2 equivalent, the difference being as much as 1 ton of CO2 per year per 1000 sq ft of a hot roof in a sunny climate. There will be many other factors that must be considered.

One could introduce an AI tool at this stage to determine the carbon cost of a building. With the correct training, such a tool would provide an accurate environment impact rating. As often happens with AI, the problem is in obtaining the relevant data. Watching energy usage over a particular period is only part of the equation; AI would be very good at that. Obtaining accurate data for actual emissions would be problematic in the overall sense of the building. The AI tool for determining energy usage for a structure would be a valuable part of the LCCC calculation.

The most effective decisions about sustainable design are made early in the design process. Traditionally, energy and performance analysis is made after the architectural and design documents have been completed making it very difficult to modify the design retroactively. BIM provides information about the building form, materials, context and the different systems to be implemented at an early stage, allowing sustainability and performance analysis to be made throughout the process. Having the extra carbon views associated will produce a more realistic true carbon cost.

Using the current tools available has lead to improvements in Carbon performance of structures. They are an excellent starting point for the designer.

However, there is a caveat, in particular, the performance gap. Performance analysis is rarely performed in depth, according to "Evaluating Optional Energy Performance"[7] low energy buildings usually use more energy than the designers expected. The UK Government view is that *"The performance of low energy designs is often little better or sometimes worse, than that of an older*

---

7   "Evaluating energy performance in buildings" National Energy Foundation 2017

building they have replaced, or supplemented" which has become known as the performance gap.[8]

Another reason for the performance gap is because engineering systems must be commissioned effectively, and the operators and occupiers must understand how to operate them. Unfortunately, the different stakeholders involved during the different stages of the design and construction of the building have different economic interests in terms of investing in energy efficiency procedures. The long term running costs of which energy is 40-80% are not usually the primary concern of the stakeholders.

SUSTAINABLE DESIGN

One way to reduce GHG emissions is to make the buildings last longer, effectively spreading the cost of carbon over many years. This is achieved through sustainable design.

The Principles of Sustainable Design are, according to McLennan "the design philosophy that seeks to maximise the quality of the built environment, while minimising or eliminating negative impact to the natural environment." McLennan[9] breaks this down into basic principles.

- Learning from natural systems (Bio-mimicry Principle)
- Respect for energy and natural resources (Conservation Principle)
- Respect for People (Human Vitality principle)
- Respect for Place (Ecosystem Principle)
- Respect for the future ("Seven Generation" Principle)
- Systems Thinking (Holistic Principle)

McLennan suggests that it is not just about ticking boxes and improving design piece by piece; important, but there is a bigger picture. Here we discuss this bigger picture, staying away from items that are easily quantifiable.

Buildings should improve the quality of life and wellbeing of those living in and around them, who are more comfortable and healthier. Using Winston Churchill's words, "We shape the environment and then It shapes us"; form is an important part of sustainability. The carbon cost of constructing the Roman Pantheon is mitigated over two thousand years to almost nothing, so form and durability are important.

The building needs to reflect the environment in which it is to be placed. Orientation and materials must be selected to maximise the healing effects in cold climates and vice versa. Designing a building that has large areas of glass facing the sun in a hot climate will guarantee a large energy cost for cooling. Solar design principles are basic.

Other climatic features should be taken into account such as wind direction and strength, flood plain issues, hurricanes, forest fires, and so on. All have an impact of one sort or another. It is difficult to see how some of these will be built into the "carbon view"; perhaps a statistical model is needed.

Building upon virgin land has an impact that must be considered. Every tree felled has a long term effect on the carbon balance. Trees are carbon filters, sucking CO_2 out of the air. 95% of the carbon in a tree is from the atmosphere. Grasslands, wetlands and so on are fairly well protected but still must be considered in the site selection. When demolishing, we owe it to subsequent generations to be restorative.

Waste comes in many forms. Sometimes we consume more just because we can. Although only being 5% of the world's population, the USA consumes 25% of its oil, 23% of its coal, and was the top emitter of CO_2, and still is if you ignore Gibraltar (to do with bunkering fuel for ships?). It consumes more electricity than everyone else except China, which has nearly five times as many people.

To state an undeniable fact, the aesthetics of buildings are important, especially when concerned with longevity. Concerns for improving the human condition, both physical and emotional health, of improving the quality and balance between work and play, are often defined through the quality, aesthetics and performance of buildings. Their beauty can directly influence their longevity or life of a building. For example, listed buildings will be kept forever, and the ugly sixties housing ideally would all be demolished as soon as possible.

Green buildings are neither ugly nor beautiful; they are buildings that have been defined by a different process with another set of aims. Long-lived buildings have an embodied carbon cost that is small on an annual basis, but that is not always obvious to the beholder.

McLennan's[10] last principle discussed in the chapter "Respect for Process-The Holistic Thinking Principle" states that in order to get

8   "The gap between predicted and measured energy performance of buildings: A framework for investigation", De Wilde, P., Automation in Construction, 2014, Elsevier

9   "The Philosophy of Sustainable Design" McLennan, J., 2006

10   "The Philosophy of Sustainable Design", Jason McLennan, J., 2006

a better result in the sustainable design process we must change the design process. He states the basic principles as;

- *"A Commitment to Collaboration and Interdisciplinary Communication*
- *A Commitment to Holistic Thinking*
- *A Commitment to Life-long Learning and Continual Improvement*
- *A Commitment to Challenging Rules of Thumb*
- *A Commitment to Allowing for Time to make Good Decisions*
- *A Commitment to Rewarding Innovation"*

In analysing these points, we see this is the perfect entry for BIM, where most, if not all of these principles are inherent in every system. Furthermore, McLennan states "Sustainable Design requires a commitment to breaking down the barriers between the disciplines," which BIM already does. BIM coordinates the diverse information required to do this. Sharing this information, together with the mixed experiences and knowledge, should bring better solutions.

Quoting McLennan again, successful sustainable design requires a shift *"in status quo thinking about how things are put together, how they are operated and how they are maintained. If sustainable design had a single name it would be holistic thinking"*. These embody BIM. BIM involves these exact process changes. One of the important features of BIM is that the designer can assemble components into the virtual building into which can be built a solution to a series of design problems, an awareness of the assembly and maintenance aspects and specialised design abilities. The components can also take in information from the real world in real-time, making this awareness in sustainability terms very powerful. Holistic thinking connects all things, the design with the environment; it asks questions as to why and why not, it searches for new materials and technologies to break down the barriers.

BIM components should allow the embodiment of knowledge and progressive refinement of the elements making up the building.

One of McLennan's[11] points is that rewarding innovation is important. In the past, master stonemasons carved their initials onto great works, painters sign their works, modern game-code writers get their names on the credits, and of course, great cooks have dishes named after them. Perhaps like the credits in a movie, BIM components embodying exceptional innovation (and they do exist) need a view containing who did what and applauding it.

Sustainable design is typically long-lived, allowing the world around to endure. A sustainable design complies with the principles of social, economic and ecological sustainability and so such a design is a design that allows the building, its users and the world around it to flourish. In summary:
- Low impact materials, non-toxic, with low carbon impact in terms of production and transport
- Energy efficiency
- Longevity
- Reuse of materials
- Comfortable, inspiring, uplifting
- Psychologically comfortable including user control, connection to light, boundaries and beauty

SUSTAINABLE URBAN DEVELOPMENT

There are hundreds of millions of small residential dwellings, and each represents the opportunity for a small reduction in GHG emissions. Housing accounts for a significant part of CI GHG contribution, somewhat over half. Unlike other sectors such as energy generation, the building sector has a large number of sites with a smaller opportunity. It is "relatively easy" to achieve a reduction in GHG when you have a single power station, but much more difficult if you have millions of homes.

The same criteria need to be applied to smaller dwellings. It is an important part. How this is achieved is problematic, but it has been shown that fine tuning the energy consumption of buildings can have significant savings of up to 40% on energy usage.[12]

Building design is the result of many factors being melded together, where adding a set of constraints regarding sustainability creates new challenges. Knowing this, together with previous sections on design principles in mind, we can look at which of the properties of buildings are most quantifiable.

- Day lighting
- Air Quality
- Passive Solar Heating
- Natural Ventilation
- Embodied Energy of materials
- Energy Efficiency
- Water Conservation
- Commissioning
- Solid Waste Management
- Renewable Energy

11 "The Philosophy of Sustainable Design", Jason McLennan, J., 2006

12 "Zero Energy Buildings: A Critical Look at the Definition", Torcellini, Pless, Dero, and Crawley, Conference Paper, ACEEE Summer Study Efficiency in Buildings, 2006

- Natural Landscaping
- Site Preservation

BIM has been used for various aspects of building design. Part of the "ultimate solution" to the sustainable building design problem in BIM terms would be a building model that gives a precise set of "sustainability" figures for all buildings for all sites.

How could BIM systems be improved to achieve a reduction in wasted resources, a reduction in carbon emissions, the costs of running and decommissioning a building? Again, we would extend the idea of "views" of each component to have a new view that pointed to functionality associated with BREEAM and LEED classification "parametrics." Each view would have a piece of the parametric language code associated with the system, or perhaps C++ or another standard language is linked into the system at runtime.

Mark Bew, co-chair of the team that authored the strategy paper, explained why improved construction data is essential: *"The strategy is all about using BIM and re-usable information to enable cost and carbon reduction."*

Learning from these issues and with the integrated building model at our fingertips, an improved way of carbon efficient design must be achieved where import and export of data is avoided, where certified BIM enforces checks and rates ALL buildings, where the BIM model allows building use and even city wide simulation, CIM-CSM, together with climate in terms of energy use and carbon generation.

The building envelope is the interface between the interior and the outdoor environment. Minimising heat transfer through the building envelope is crucial for reducing the need for space heating or cooling. Insulation, air-sealing and windows play an important role in heat transfer, but at present, these are almost the only design criteria engineers can calculate. BIM offers so much more.

THE GLOBAL IMPACT OF BIM

The global construction market has passed $17 Trillion ($1,700 billion) in 2019, representing perhaps 20% of the world economy. This industry serves "almost all other industries", and so is a significant part in the global scheme of things.

The Boston Consulting Group have concluded that within ten years *"full-scale digitalization in nonresidential construction will lead to annual global cost savings of $0.7 trillion to $1.2 trillion (13% to 21%) in the engineering and construction phases*

Figure 223. *Computer Generated Images © 345 Holdings 2019*

and $0.3 trillion to $0.5 trillion (10% to 17%) in the operations phase." Digitalization will change the game fundamentally in E&C [Engineering and Construction], not only enabling efficiency and quality gains along the value chain but also reshuffling the competitive league table of companies and countries."[13]

These are immense savings. It has been estimated that $500 Bn could solve world poverty.

The UK Government seeks a 50% reduction in Greenhouse Gas (GHG) emissions through digitization and smart initiatives in the construction industry, much of this is based on BIM. This sector contributes 30% of the GHG emissions and consumes up to 40% of all energy[14]. Should this ambitious aim be met the industry could

reduce emissions by as much as 20% or up to 700 million metric tonnes of CO2 equivalent. The long life of buildings means that changes made will be utilized into the future. The huge projected growth (70% to 2025) in the Construction Market means that it is crucial that the GHG not be allowed to grow with this increase in activity.

The main difference, of course, between the economic and sustainable is that the economic changes will happen simply because the economics are the driving force while sustainability will need to be pushed. Providing the right tools will help both causes.

According to the Government white paper Construction 2025: *"The revolution does not stop there. It brings us buildings and infrastructure that are more usable, better looking, more efficient, better managed and safer."*

BIM and its derivatives will have a huge impact into the future on the world's economy, on GHG emissions, on city living, and will impact the lives of every citizen.

---

13 "The Transformative Power of Building Information Modeling", Gerbert, Catangnino Rothballer, Renz and Filitz, Boston Consulting Group, 2016

14 "Buildings and Climate Change", UNEP, Sustainable Buildings and Climate Change, 2009

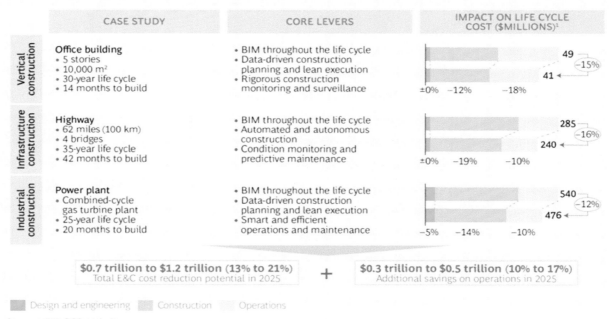

EXHIBIT 3 | A Digital Transformation in E&C Could Reduce Annual Costs by More Than $1 Trillion

$0.7 trillion to $1.2 trillion (13% to 21%)
Total E&C cost reduction potential in 2025

+

$0.3 trillion to $0.5 trillion (10% to 17%)
Additional savings on operations in 2025

Design and engineering   Construction   Operations

Sources: IHS; BCG analysis.
[1]Life cycle cost is shown as the inflation-adjusted net present value. The cost of equipment (for instance, gas turbines in the power plant) and of non-building-related operations (for example, fuel for the gas turbines) is excluded.

Figure 224. *BIM Global Savings. The Transformative Power of BIM 2016 © Boston Consulting Group (BCG)*

Conclusion
16

It has been evident for many years that the design, construction and management of buildings has been a complicated and inefficient process, achieving less than optimal results. Even the advent of computers had not helped, with improvements in various aspects, but not in the overall process. The construction industry has lagged behind perhaps all other sectors in its bid to become truly digital. BIM has given the industry a push in the right direction, with many digital applications following in its wake. In many countries, BIM has reached a critical momentum, with Governments and their Construction Industries recognizing that these benefits can be achieved.

Early systems have attempted to solve the issues of designing and building. BDS (Cambridge UK) provided a weather vane geometry that ensures the coordination of 2D drawings. This concept has been carried through to SONATA, REfLEX, ProReflex, ArchiCAD and Revit. SONATA introduced working on plan elevation and 3D windows, icons and pull-down menus into the BIM space. It combined this with user-defined parametrics objects, 3D shapes, objects and parametrics, automatic wall and window joins including cavity closures in all dimensions giving us BIM as we know it. SONATA users were the first to share the Single Mode. The authors of ArchiCAD have graciously acknowledged that SONATA was the first BIM system in 1985 (appendix 6).

Building a computer-based model of the structure with all its detail brings an understanding and a certainty to the many aspects of the design. It brings a host of efficiencies, including collaboration and has led into a wide range of procedural, work practice and legal changes that have streamlined the industry. It has achieved, significant savings in cost, time, effort and improved sustainability and quality have been achieved.

Enormous savings are being made, and in 2025, this is estimated at $1.7 Trillion per annum World-Wide. According to the British Government, BIM significantly reduce GHG emissions potentially by as much as 50%.

New terminology and in fact, a new industry and technology has grown around BIM. Potential users must learn new systems that are complicated and involve a different way of thinking. No longer is the primary output drawings, but rather a complete model from which all information is extracted.

The historical case studies in the book show how many of the BIM ideas were introduced across many different disciplines. The wide-ranging benefits of BIM are enumerated in the majority of disciplines associated with the Construction Industry.

Retail Information Modelling has been investigated, partly to illustrate where BIM can go, but also in its own right. Using RIM, we have found exciting new design environments using BIM concepts of information and dynamic interaction, but adding Artificial Intelligence and new technologies, giving us new interfaces and insights unseen before. The seeds of improving, expanding and enriching city life, using City Information Modelling, have been sown.

With the advent of Artificial Intelligence, we will soon see the automatic generation of building designs, maintenance, costing, as well as optimization of buildings and costs. This bodes well for quality of life and managing an increasingly complex world. Climate Change brings a series of very difficult challenges we must rise to meet.

Appendix 1
Theory and Background: SONATA

This appendix describes some of the tools developed by the Author to implement SONATA. A complete list of the tools that were written is included in Figure 225.

This information is included to give an insight as to the history, the issues and hopefully an understanding of the underpinnings of BIM.

Three-dimensional visualization is an important part of the design process for almost all structures. Visual impact, accessibility and even comprehension of the overall design is achieved through 3D images. It allows non-professionals to examine buildings, check lighting and shadows, each and all essential within the integrated model. The Author developed various algorithms over 20 years; almost all the images and hidden line drawings in this book are from this work. The Author extended the algorithms to include

shadows, smooth shading, fog effects, transparency, textures and spotlights, and perhaps most importantly, to be efficient within computing constraints of the time. Many figures here show these effects.

The earlier images were produced with 1MHz processors, 0.1MB RAM. This constraint meant that considerable effort went into making the various parts of the algorithm efficient. The advent of GPUs and game engines have made the efficiency issues irrelevant but it is useful to know some of the processes involved in generating these images. We take for granted the live realistic scenes that appear in games and AR.

Systems produced a series of flat faces defined by a series of three-dimensional coordinates, with neither particular rhyme nor reason to the order of the faces. These faces must be rotated and

Hatching	Parametrics 2D/3D	Boolean 3D operations
Fonts definitions & plotting	Self-solving Networks	Closure detail cavity walls
2D shapes	Database structure	Clash detection
Snap Codes Line/Arc/Spline	User defined objects	3D shapes
Rendering algorithms	Language definition	Icons drawing
Hidden Line algorithms	Geometric Projections	Pull down menus
Spot Light	Clipping to Frustrum	Anti-aliasing
Fog	Box Geometry handling	Multiple project access **
Shadows	Flight path for videos	Icon file formats
Reverse Polish expression	Libraries of objects**	Plotting software
Drawings/Sheet handling	Wall Join Details	Polish Code executor
Bitmap generation	Window handling	Overall interface design

Figure 225. Tasks completed for SONATA 1997 (** after 1987)

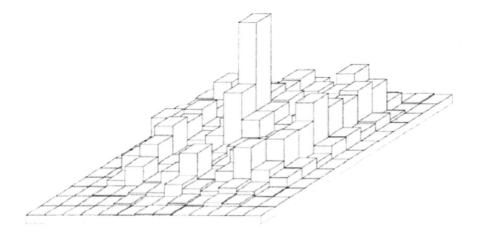

Figure 226. *3D representation of eyeball movement following letters. 1975.[1]*

1   Original Data from Dr. Danny Latimer, University of Sydney

projected onto a notional screen in front of the virtual viewer. Faces outside the view must be clipped away. The process then starts working out which bits of the faces (model) would be visible and which would not. For example, the objects on the other side of a solid wall are not visible. One has to know how far away every point on every surface is (for each pixel) and whether something is closer. The closest surface is drawn with a colour determined by the lights, shadows, surface type and orientation.

This algorithm took an edge representation of plane facets, projecting data onto screen coordinates, and sorted the data with a bucket sort (a linear growth technique), one bucket per scan line. Scanning was from top to bottom of the scan line, maintaining an active edge and facet list. Edges were sorted using a bubble sort. The distance from the eye to each point on the facet was calculated and saved into a single dimensional array (an incremental calculation). Shading was calculated on a facet-by-facet basis and assigned colour values added to each visible pixel. This was called a single-dimensional Z-buffer. Even choosing the edge representations type, for which there were alternatives, involved guesswork. The lack of "core" memory (and it really was magnetic cores on wire thread) meant the algorithm had to be something more sophisticated than today's frame buffers, certainly in terms of memory use. This algorithm was used by SONATA and REfLEX (possibly other well known CAD systems) to produce most of the images in this book.

The images (Figure 230) of the ball and tori was on the cover of the Division of Computing Research magazine in November 1975 and is an example of the evolving one-dimensional Z-buffer algorithm. Cyan magenta yellow (CMYK) versions of the images had to be printed separately, to get the colour image. There were no colour monitors available at the time. Single-line Z-buffer techniques had become the main method for generating images. This technique used was again to "bucket sort" vertices in projection space and then scan from top to bottom of the "screen". Edge lists were maintained with edge crossings (potential changes in the visibility of a line or edge) calculated exactly and stored. When edges existed, the current scan was processed and output: or not if they were not visible. This algorithm evolved over the years into something more accurate and faster. Floating-point calculations were counted to determine exact efficiency.

Using this technique made something like transparency easy; just another line of Z-buffer depths for the object behind the transparent object. The colour of the pixel is simply a summation of two-pixel colours, depending upon transparency and angle of the front face. Similarly, fog was relatively easy to simulate and allowed images to have a somewhat ethereal effect (see the images in the shadows chapter).

Textures, completed somewhat later in SONATA, are again straightforward if you ignore aliasing: however, where aliasing is ignored, the images were appalling! Something called summed texture maps won in the end, though that was never obvious. Generating shadows was complex, involving generating hidden-line views of the scene from the light source viewpoint, reconstructing the visible faces and adding them to the scene.

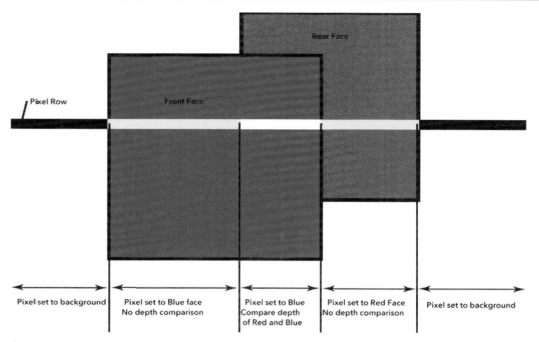

Figure 227. *Simple Scan-Line Algorithm showing faces as projected onto screen.*

(see Figures 232 and 233). Internal scenes especially looked totally bland. To help add feeling to the images, the Author added spotlights. The user could define a colour, a range, and an intensity cone. The light intensity cone was defined by points to which a spline was fitted. This meany that a spot light could be defined with variable intensity away from the main direction of the light. There was a long-term bug that the light-cone would get a band in it that would not go away! See the bands in Figures 80, 81, 83 and 84.

It is often useful to be able to change colour within the finished images of particular objects. One can pick the colour on a particular object in a 3D rendered view and change all shades of the colour on the particular object within the view. Colour values were then changed for all instances of the component in the image by clicking on the buttons on the bottom of the screen. Figure 76 shows the office with colours changed. The buttons on the bottom of the screen allowed the selected object to change the colour via Red, Green and Blue, Hue Saturation and Intensity, and the Ambient Light.

Hidden line algorithms were always a challenge because of the large numbers of edges generated from full production drawings and the need for complete accuracy. "Pretty" 3D hidden line

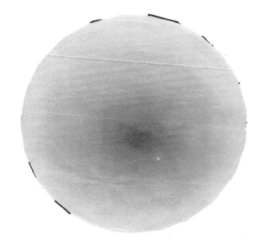

Figure 228. *First smooth shading attempt. c. 1975 © RIBA Archive, Victoria & Albert Museum*

images were less critical in terms of construction implications than the drawings, so a variant on the hidden line algorithm used for the 1976 movie was used. In the end, it worked well, but there were numerous issues on the way. In some ways, the calculations were more complex than the rendering algorithms.

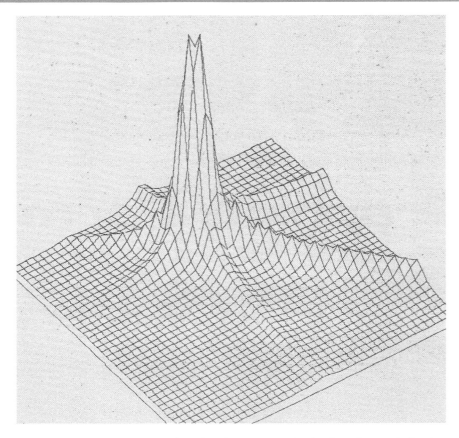

Figure 229. *Hybrid hidden line algorithm displaying amplitude of machine tool vibration versus input frequencies 1975*

In 1976, an opportunity arose for the Author to make a 3D building model and view it as a movie. Several architects proposed new designs for the Commonwealth Courts in Hobart, Australia and the Division of Computing Research had a new 16mm camera mounted on a high-resolution graphics screen (III Comp 80). It was impossible to make an image version of the movie because the massive (for those days) amounts of data generated, but a hidden line movie was feasible, from which a single image from the data is included as Figure 259. The movie itself is included as part of the UnderstandingBIM AR app available at app stores.

The movie credits show "Architects: Peter Fox and John Hodge" and "Technical Engineer: Jonathan Ingram" and there is a single frame in the movie between the opening credits and the actual street scene, that is a "billing frame" dating the movie to November 7, 1977. This movie is probably the first of its kind, a 3D movie of a real architectural scene. The original 16mm film hard copy is with the RIBA archive, at the Victoria and Albert Museum in London.[1]

A virtual model of the street scene was constructed requiring data-structures. The facets making up the objects made up from manifold (closed) shapes and were constructed with boundary representation (B-rep) data structure. The internal data structure to represents the facets and objects were not clear. After some trial and error, four separate lists of vertices, facet edges defined by pointers to the vertices and facets defined by pointers to the edges and objects by pointers to the facets, were chosen. These objects were saved as separate named files representing different parts of the buildings. Faces were defined clockwise; those facing away from the viewer were removed quickly with a dot product test normal to the face.

Camera paths and the viewing direction were generated by using a 3D spline through hand-generated points. This algorithm was

---

1   RIBA Library Drawings & Archive Collection, Victoria & Albert Museum

Figure 230. *Author's first colour image separate CMY layers 1976 © RIBA Archive, Victoria & Albert Museum*

written on a coding pad in FORTRAN, and transferred to punched cards by punch card operators. This was submitted to Cyber 76 to generate the vector to be recorded on the 16mm sprocketed camera. Odd artefacts occur in the movie because of some edges sitting almost exactly on other faces. Sometimes, the calculation would place the edge in front of the face, sometimes behind.

This early work on hidden line and surface algorithms, building representation, database, 3D interfaces and movie generation,

was pivotal in SONATA and REfLEX work. All of the SONATA images and some REfLEX images were produced directly from these early algorithms.

The idea of treating building components as objects was central to the SONATA development. Stroustrup[2] had introduced the idea at a programming level with his ideas of Objects on C++. In the C++

2   *"The C++ Programming Language"* Bjarne Stroustrup, Addison Wesley, 2013

Figure 231. Colour separations used to make the previous image 1975 © RIBA Archive, Victoria & Albert Museum

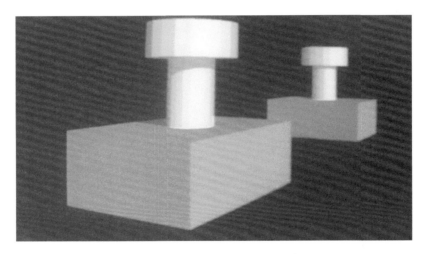

Figure 232. 1984 Hidden Surface view with shadows © RIBA Archive, Victoria & Albert Museum

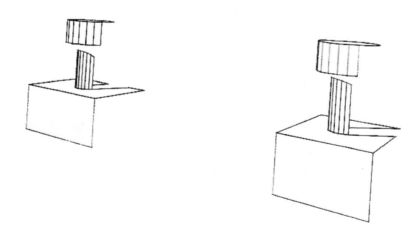

Figure 233. 1984 Hidden Line for shadow generation. © RIBA Archive, Victoria & Albert Museum

Object Orientated computer programming language, programmers are encouraged to think of objects which come together to make a program as stand-alone units that contain data processes and an ability to communicate with other objects. They contained their own data, and they contained only methods that were relevant to that object. REfLEX and SONATA both used this concept.

While working at the CSIRO in 1987, there came a need to build an interpreter to analyze, store and execute arithmetic expressions and interpret commands associated with the III COMp80 graphics device. The Author designed and implemented a syntax analyzer to break down expressions and commands into their individual parts: then expressions were executed using a technique called Reverse Polish Notation (RPN). It involved pushing values and expressions onto a first-in last-out stack and solving the expression by popping off into registers when done. This worked well and was reasonably straightforward. This piece of code also did the overall command syntax check.

The resulting interpreter simulated the complex multi-million dollar COMp80 machine on a Tektronix storage tube of the time. It analyzed commands and executed them; drawing a line, outputting some text, changing font, evaluating an expression and so on. The purpose of this interpreter was to see the results of typesetting pages without having to go to the COMp80. The MIMS journal was set using this. This was useful code when writing SONATA, as it slotted straight in, albeit with some different commands. Appendix 2 shows the full range of commands implemented with SONATA compiler/interpreter. Many parametrics were written for SONATA, from simple windows and doors to groundworks for roads and buildings, to structural analysis and HVAC network solutions. The chapter on Engineering BIM illustrates some of the applications.

SONATA parametric components were saved into the database as "compiled" expressions and as ASCII text. The expressions included mathematical functions such as SIN and COS as well as variables from within the element. The information stored in the database was assembled lists of values, operators, functions and variables ready for direct execution, and also the text equivalent of the expression. The list of values and operators in the string were interpreted and expanded by SONATA at run-time. SONATA stored the stored ASCII text to allow the user to edit and "recompile" the parametric text at a later date.

SONATA had to have its own compiler to break down mathematical expressions and different commands in a specifically designed language.

REfLEX was designed so that components were "cross-compiled" into C++ objects, a single BIM component being represented by a C++ object. These were then compiled using a regular C++ compiler and bound into the Dynamic Linked Libraries (DLL) to be available at run-time for REfLEX. Whole libraries of components were built into these DLLs.

A seamless merger between C++ and REfLEX (and subsequently Revit) objects occurs, enabling the use of C++ characteristics such as Abstraction, Encapsulation, Inheritance, Polymorphism and Overloading. When you define a class or a component, you define a blueprint for an object. SONATA had been written in Fortran before DLLs were available. This is, of course, is directly related to the Revit idea of families.

Commands available within the parametric were diverse. For instance, components could move themselves, be aware of the passage of time for moving and changing, reference external databases and sensors, pass values to each other to solve engineering problems, appear differently depending upon viewing information such as scale or distance to the eye, scan building layers and libraries to evaluate areas, and so on. As an example, any components could contain a scan command as part of the parametric syntax to scan for particular or groups of components. A "scan fittings layer selection view" would loop over all elements in the fittings layer, selecting based on a criteria defined in a selection (also stored as an object) and restricted to a particular view area in the building defined by the view object. This enabled objects on a drawing to print a schedule or an air vent to determine where the surrounding walls were, for example, albeit rather slowly.

Various other reporting mechanisms were available. Components could refer to complete files of information <<filename>> and also contain an execute a command for another component. <<<element>>>. When found, the element would be executed. This was exclusively used for printing schedules, though there was no reason to stop other more intricate calculations being performed.

In 1983 a solid modelling program, probably Unisolid, was being demonstrated, at a conference attended by the Author. Here shapes were cut from each other and added together in an imposing way. It was apparent that SONATA would need something similar to generate the 3D shapes and allow components to cut holes in each other. It was an obvious solution to the need for windows to cut holes in walls and to cut voids in slabs in models, both in terms of calculations, lighting and visualization. During that year, the Author wrote code to do these "Boolean" operations longhand, onto coding pads(!). This code

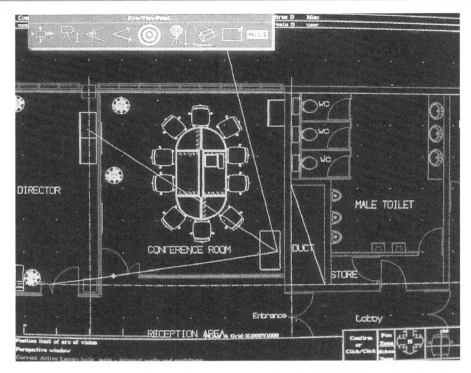

Figure 234. *SONATA Interface: positioning a perspective view © RIBA Library Drawings & Archive Collection, Victoria & Albert Museum*

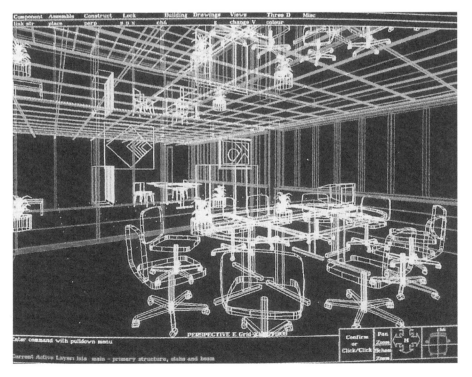

Figure 235. *SONATA Single Perspective View generated from Figure 234 © RIBA Archive, Victoria & Albert Museum*

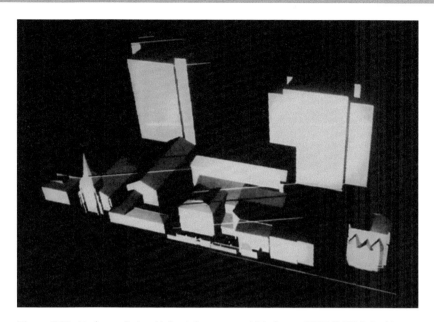

Figure 236. *Early rendering Hobart Commonwealth Court 1977 © RIBA Archive, Victoria & Albert Museum*

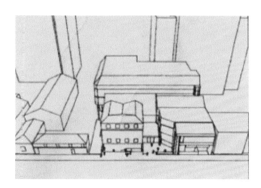 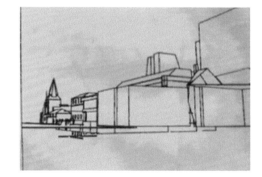

Figure 237. *Hobart Commonwealth Court plotted on Tektronix Tube printer © RIBA Archive, Victoria & Albert Museum*

sat for a year and was then typed into the WCW (Whitechapel Computer Works - see below) and debugged as part of SONATA parametrics, object linking and clash detection.

The first workstation (WCW) used for the development of SONATA did not have a line drawing capability. There was no graphics capability at all. (See Figures 241 and 242.)

While supervising the Author's PhD, Professor Pitteway from Brunel University helped with implementation of grey-scale version Bresenham's line algorithm to draw a straight line between points on a grey-scale pixelated screen. One might think this

straightforward, to draw straight lines, but it was surprisingly difficult at the time. Drawing curves increased complexity. The eye is very good at noticing inconsistencies and lack of symmetry, where problems with the algorithm and resulting graphics were creating a smooth line, straight or curved, without doing divisions and multiplications at each pixel on the line. The lines were smoothed using an extension to Bresenham's algorithm. As stated elsewhere, the 1984 WCW ran rather slower than today's (2020) machines, so every repeated calculation was critical to the performance of the system. Floating-point calculations were again counted!

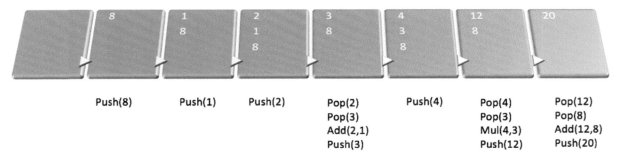

Push(8)   Push(1)   Push(2)

Pop(2)
Pop(3)
Add(2,1)
Push(3)

Push(4)

Pop(4)
Pop(3)
Mul(4,3)
Push(12)

Pop(12)
Pop(8)
Add(12,8)
Push(20)

Figure 238. *Reverse Polish Stack for solving 8+(1+2)*4*

Window handling is taken for granted with one window appearing over another naturally and without effort in modern systems. Pull down menus and temporary windows are all examples of this. The WCW had no such capability. Programmers had no access to window handling in the machine. If windows were drawn from the back to the front, then the correct parts of each window were drawn too slowly, so window-handling code had to be written, turning out as a few lines when implemented as a recursive routine. (A recursive routine calls itself from within itself, which calls itself from within that, repeatedly until the complex problem had been solved.)

```
## filename facet.pl
##
qp "Second point",a,b,3000mm,3000mm
qp "Third point",c,d,3000mm,0mm
ql "Second height",e,100mm
ql "Thrid height",f,0mm
ql "Contour interval",g,1000mm
##
lo△1
    d0a△a△b
    d0a△c-a△d-b
    d0a△-c△-d
le
ob△(e+f)/3
##
U=1m
mx=(a+c)/3
my=(b+d)/3
mz=(e+f)/3+HEIGHT
ax=((d*e)-(b*f))/(2*u*u)
ay=((f*a)-(e*c))/(2*u*u)
az=((c*b)-(a*d))/(2*u*u)
T=sqrt(ax*ax+ay*ay+az*az)
V=mz*az/U
xc1=0
```

Figure 239. *SONATA parametric code fragment for generating contours*

When the window handling had been completed, the menus and icon handling, display lists and projection/clipping capabilities then all had to be constructed. Sometime in the early eighties, the Author saw the Rank Xerox paper detailing the mouse, icon, pull-down menu trilogy documented in 1979. The idea of using icons and menus, saving "real estate", was appealing, so the SONATA interface was redesigned along similar lines. A thin menu command line across the top, a settings line immediately below, the current command and component windows along the bottom of the screen were added. Traditional systems used the whole screen for each menu step, menu items were numbered and the numbers keyed in to select. Icons were drawn by hand, written into memory as bit-patterns, and drawn with the recursive screen handler. They were stored in SONATA's own icon format.

Pull-down menus were then constructed in areas of memory. This involved writing system text, icons and lines into specific areas of memory as bit patterns. These areas were then drawn using the recursive window handler described above. When the mouse (a new invention) was clicked on the screen, an interrupt was generated giving the overall screen coordinates. This mapped back to individual window coordinates which were mapped back to building space. Tedious is an understatement.

The double-page view of the screen of SONATA is shown in Figure 36.

In order to draw the building projected onto the overall screen space, 4x4 matrices of transformations were stacked and multiplied together, applied to the building coordinates and drawn onto the appropriate window memory, then displayed with the recursive display handler. This meant that 3D views appeared in perspective, plans as orthogonal views and so forth, on the correct part of the screen. The cursor, and any object being dragged or placed, were redrawn dynamically.

When dragging or clicking in a building window, the program obviously needed to know the building coordinate (3D). This was

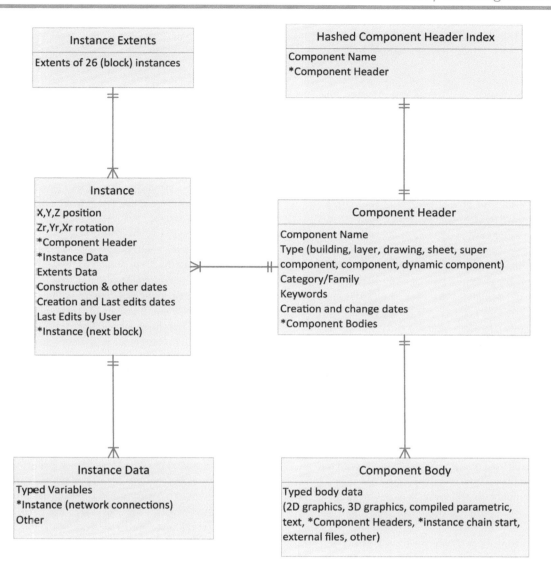

Figure 240. *Fundamental BIM data structure*

done by inverting the matrix (writing the code from scratch) and applying that to the screen coordinate (2D) returned from the click. In the case of a click on a 3D window, calculating a "ray" from the viewer through the clicked point into the building space. This ray was used to find the nearest intersection or point on ground plane or whatever. It was complicated to keep track of all the parts, but there was no need to communicate any of these programming structures to anyone else, as there was no-one else.

The system must decide which view of a component to draw depending upon the viewer, the scale, the type of view, the final

type of output, whether it is plan, elevation, perspective or 3D, and the orientation of the component itself. There is a set of rules for this which SONATA, REfLEX and Revit all share, but it's not always straightforward.

The display list handlers then had to be written. A display list is a list of graphical commands for each window that defines what needs to be drawn each time the vuport is refreshed. Each separate "vuport" needed separate display lists, and each display list had its own projection matrices to get the plan, elevation and perspective effects, which all needed to be clipped depending upon the view and projection type. View and projections were stored and named as part of the database. (These were also used for drawing definitions etc.).

The display list referenced a very troublesome function which decided which view of the element to draw. For instance, drawing a symbolic view into a window, one would expect to draw the symbolic plan view of the component. If there was not one such view then you would draw the 3D plan view, or if the item was rotated then an elevation view would be used or a 3D view if rotated through a non-orthogonal angle. There were a number of other variants to do with handing of components, changing parameters of parametric components in the display list and operating "live" components. In order to display the various graphics, a hierarchical set of routines relating components to each other was written, to apply transforms and to display in the appropriate colour, line styles etc., which involved matrix handling routines to apply 3D coordinates to screen coordinates, etc., etc..

The homegrown code was easily replaced when OpenGL came out some years later as many of the code structures were similar. Also there were no image database storage formats at the time, so the Author wrote a (rather inefficient) format that was called the ".img" format as the .jpg format did not exist until 1991.

Display lists had pointers back into the components to make sure that as the components were moved, dragged or rotated, the views in all windows were updated. The display list components were drawn onto the area of memory and "blitted" onto the screen. It was possible to draw primitives which could be assembled into components and these were all part of the display lists.

Another issue was construction lines and picking points or components in the different windows. Much time was spent writing the nearest intersection, parallel line, nearest point of line, and so on. These were time-consuming and had to be correct: See appendix 2 for a complete list. Picking or selecting components from a vuport either individually or from a selection are/was somewhat involved, resulting in the display list handler becoming more complex.

There were no fonts available, stick or full, that could be used on both the screen and output devices. The text output on the screens and the plotter had to match. Stick fonts were recovered from "the old days" and hardwired into both the main program and the plot output software. Fonts could be scaled and transformed at will, for which an outline font was created, using the hatching code to fill it in if required. The title block in Figure

243 from a 1992 working drawing, illustrates some of the fonts that had been assembled.

Hatching of various sorts is an integral part of construction drawings, so any system producing drawings must have hatching. In BIM terms, this means hatching in plans, elevations, sections, hidden line and 3D. Hatching was bounded in various standard ways, including a seed point. Working out the smallest closed loop around a seed point when you have lines, arcs, circles and splines was fraught. Line hatching inside a shape is straightforward, but hatching with a symbol or shape that might consist of lines, arcs and circles, against a shape that might consist of lines, arcs and circles is unpleasant.

Every BIM system needs primitives and SONATA was no exception. The creation of the 2D primitives took forever. Every single piece of geometric arithmetic was created from scratch. For instance, to generate a circle defined by tangent lines defined by their endpoints, the general case had to be programmed, perhaps not that difficult, but the sheer number of different combinations and functions, and rubber-banding was time-consuming and tiresome.

Code to generate 3D shapes was also all created from scratch. Diverse ranges of shapes were defined as manifold, closed faceted objects in order to work with Boolean Operator code described elsewhere. For instance, generating an extruded shape following a line in 3D-space was messy rather than difficult, but to repeat over and over for the different primitive shapes seemed endless. To have the vertices correctly aligned in the extrusion case meant some interesting vector geometry on the faces. Making sure the extrusion rotated correctly as the linked lines progressed was a also problematic. The roof primitive was particularly difficult because of the vast number of variations.

As stated ad nauseum BIM systems need to pass information between components in order to avoid duplication and to make sure all are up-to-date at all times. For instance, a window in a wall needs to know how thick the wall is without copying that data from the wall (in case it changes). The mechanisms involve pointers between the components and components being able to request information across the link. SONATA links were done by the link command, but Revit has since automated this by linking families and classifying the components into families. What then happens in both systems is that walls cut a hole for the window (using a Boolean 3D operation) with the dimensions coming from the window, while in the symbolic views, appropriate wall closing components are inserted where the wall is broken. Change the window size, and the wall acts accordingly.

Other systems of the time, including RUCAPS and ArchiCAD (at that time), had no concept of this linking or insertion process, which doesn't end w th the breaking of the wall in plan. Another was in multi-way wall joins, dealing with cavities, matching the hatching and making sure all the different views matched.

The link functionality described in the previous section was used to form networks of components. Loops could be generated automatically with an exchange of lines for components, correct insertions and linking at corners, etc. They could also be built by linking elements together using vectors on different faces of the components to define the position and orientation of the new component. Effectively they clip together exactly as Revit does today. Network loops were treated as single components for some purposes, reflected in modern systems.

The Author had experience with various networks and was keen to apply the principles learned. Finite Element Analysis (FEA or FEM) work in the 1970s on ship hulls suggested nodes with balanced forces or flows. Solving pressure/diameter water pipe networks by hand as a student engineer inspired the use of flows between the nodes.

Within the building model, components can be linked so as to pass information to each other connect. Nodes are defined at their connection points. As stated previously, a series of linked components or networks can be made manually or from placed centre lines. Vectors are used to determine how the duct links, an up vector and a direction vector, identical to the Revit mechanism.

Similarly, at each node where the building components join, the forces, or flows must balance. Various constraints must be considered. For example, solving the pressure, temperature drop and noise levels in a HVAC duct in a building requires consideration of the incoming flows at each end (outflowing is -ve), the pressures, sizes and physical properties of the duct, the equations, building standards, possibly the environment outside the duct and so on.

Extra constraints and the complex way in which they might be applied led the Author to determine a different more general method of the solution rather than the partial differential equations of FEM. Each component in SONATA and REfLEX included a physics "view" (a view is a particular way of looking at a component), the constraints, equations, standards, resizing capabilities, 3D "awareness" to allow itself to solve the full design problem within the model.

Using a straight simple HVAC duct as an example, the flows into a duct must sum to 0 (out is -ve), the sum of the pressures must equal the pressure drop for the flow and properties. The component must resize itself to stay within the standards. The adjacent components must also be "informed" of the change in size, so they match in terms of size. Early versions had the sizing done as a separate process, but the resizing affects the flow and pressure, so it must be done as an integral calculation. The calculations in the physics view of components had the remarkable property of the order in which they were performed irrelevant!

The same process must be for other pieces in the network. A "T" piece had three in-outs to consider and 3 connection points. A simple nozzle had one end connected to the duct network but the other end going into a room or hall. This required certain considerations, but once you have a network defined with all the constraints and physics views defined, the network can be solved. This was done by repeatedly "executing" the physics views of the components, not in any particular order until the network ceased to vary.

An example is given elsewhere (Chapter 9 BIM & Engineering) regarding Huddersfield Football Club Stadium, where similar principles were used to consider options for potential sites around the town, redrawing the whole stadium for different numbers of spectators and pitch sizes.

Note: In order to write this amount of code I set myself a daily target. There was no Instagram, Facebook or Linkedin. There were no emails or texts and there was no one to talk to in the professional sense. Just me and my machine. Murray arrived several years later. He worked mostly at night testing code I had written in the day.

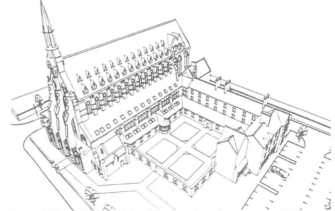

Figure 241. *SONATA Hidden Line Drawing, from Sonata RUN magazine 1989 © SONATA Systems*

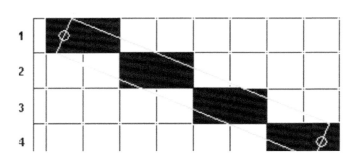

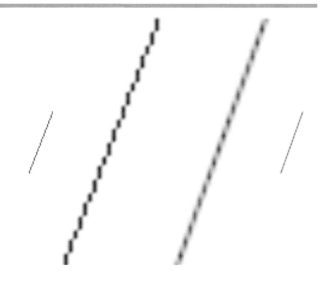

Figure 242. *Jagged edge created when using simple Bresenham's Algorithm*

Figure 243. *Jaggy compared to anti-aliased lines*

Figure 244. *First CGI by Author on Tektronix tube 1974 © RIBA Archive , Victoria & Albert Museum*

Figure 245. *Rietveld Chair. Shadow Study 1988 © RIBA Archive, Victoria & Albert Museum*

Figure 246. *SONATA Title block showing different fonts, line and filled, hollow and special 1992 © RIBA Archive, Victoria & Albert Museum*

Appendix 2
SONATA Reference Manual

The SONATA reference manuals run to some 490 pages, so only a few examples have been included in the appendix, enough to show basic BIM functionality.

Basic commands in SONATA have already been discussed in an earlier section and will not be elaborated here. Figure 247 shows the menu list of commands available in SONATA

Element menu
- Lines
- Link Lines
- Arcs
- Circles
- Text
- Shapes
- Delete Lines
- Hatch
- Obscure
- Parameterise
- Scale Lines
- Change Length/text
- Change pen/style
- Extend lines
- Parallel lines
- Link Point
- SAVE ELEMENT
- LOAD ELEMENT
- Display Element Views
- Clear Element

Assemble menu
- Select Element
- Place Element
- Dimension
- Drag/Shift
- Rotate/Hand
- Delete
- Exchange
- Join
- Change params
- Stretch
- Matrix Place
- Radial Matrix
- Rotate 3D
- Nest Create
- Nest Burst
- Copy
- What
- Report
- Resize
- Duplicate Check

Construct menu
- Construct On/Off
- Working Grid
- Bisect points
- Parallel
- Parallel points
- Bisect Angle
- Tangent 2 Circles
- Tangent 1 Circle
- Perpendicular
- Distance/Angle/Area
- Construct Clear

SNAP menu
- Point
- Jump link point
- Intersect/end
- Point on line
- Midpoint-line
- Midpoint-points
- Intersection
- End point
- Gridpoint
- Walk
- Reference point
- Perpendicular to
- Set Distance (D)
- Centre circle/arc
- Offset along line
- Link Point
- ORTHOGONAL
- HORIZONTAL
- VERTICAL

Building menu
- Input/Change Building
- Input/Change Layer
- Activate Layer
- Select Layers
- Change User
- Copy Layer
- Limit Active Layer

Drawings menu
- Fast Plot
- Input/Change Drawing
- Input/Change Sheet
- Expand Drawing
- Input/Change Selection
- Input/Remove Request
- Background Plotting

Views menu
- Input/Change View
- SAVE VIEW
- Show View
- Single Window/Screen
- Four Windows/Screen
- Change Overview
- Remove Hidden lines
- Image
- Display Image/Hidden

Three D menu
- Create Subelement
- Combine Subelements
- Delete Subelement
- SAVE 3D ELEMENT
- LOAD 3D ELEMENT
- Input/Create Spotlight
- Input/Change Colour
- Change Colour
- Generate Spotlight Shadow
- Clash Detect
- Clear subelement

Utilities menu
- Load views from 3D
- Load views from disk
- Save views to disk
- Print
- Export Data
- Transfer Elements
- System Switches
- Change Element
- Change name
- Delete Element View
- Send Message
- Show Messages
- Database Statistics
- EXIT

Function Buttons
- Grid Switch
- Pen/Line Style
- Search
- Help
- Categories
- Text
- Scale
- Element
- Sheet
- Dimension Toggle
- Change Active Layer
- Building Elevation
- Overview
- Redefine Menu
- Display Menu
- Confirm

Glossary of Terms

Appendix 1
- Special Colours

Appendix 2
- Special Keywords

Index

Figure 247. SONATA Command Summary from Reference Manual © SONATA Software Systems

SONATA MASTER LIBRARY

The Master Library provided a catalogue of 192 elements and a range of abstract graphics such as dimensions, etc. It was not meant to be comprehensive but provided a framework from which the users could build further elements. A colour library was also provided. A few sample pages are provided here to illustrate the breadth of the component library.

SONATA came with a range of other libraries, some in-house and others were created by different customers. Many elements available in the libraries were parametric, with all views were parameterized.

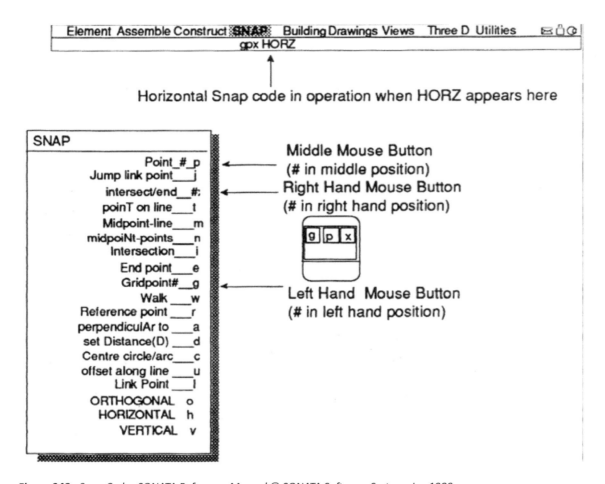

Figure 248. *Snap Codes SONATA Reference Manual © SONATA Software Systems Inc 1988*

Solid of Revolution by Fence

Operation

On picking this icon, the following prompt is displayed :

Identify first point on fence
Pick a point to position the first point of shape. Choose a suitable window for the axis of extrusion

Identify next point on fence
Continue picking points until the desired shape is obtained. Upon confirming the prompt is :

Specify axis of revolution (x,y or z)   z
The specified axis of revolution will act as a spindle which passes through the point specified below on the axis of revolution e.g. for a wine glass the axis of revolution is z and the spindle point the center of the glass.

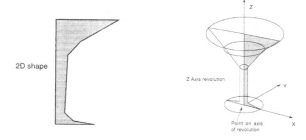

So here the 2D shape will have best been defined in an elevational window

Identify point on axis of revolution
Pick the point around which the revolve will take place, i.e., the spindle of revolution will pass through this point.

Enter number of sides around   9
[Return] for the default, or enter new number and [Return].
The number of sides around will determine the number of facets in the object. This number may be greatly reduced if the object is to be smooth shaded.

Figure 249. *Generating Solids of Revolution SONATA Reference Manual © SONATA Software Systems Inc 1988*

Figure 251. *Data Exchange, SONATA Sales Assistance © SONATA Software Systems Inc 1988*

Figure 250. *Marble Textures from Library*

3.6   Complementary Products

The following diagram illustrates the principal types of program that complement Sonata with examples of software packages that have already exchanged data with the Sonata database. The list is by no means exhaustive. Every country is likely to have its own home-grown software which may well be able to communicate with Sonata. As a general rule, if the package can format data into an ASCII, DXF or Tiff files, then it should be able to exchange information with the Sonata database.

Exchange Format

Briefing data/	ASCII	------------>	
Facility Management Databases			
eg Oracle			
D/base V			
Lotus			
Other design/drafting systems	2D/3D DXF	-------------->	
eg AutoCAD[1]			
Alias Upfront			Sonata
Moss (civil engineering)			Database
Building Analysis	ASCII	------------>	
eg Facet (M & E)[2]		<------------	
LEAP 5 (structures)			
Reporting/Presentation Tools	DXF, PICT,	<------------	
eg Word processing	TIFF, HPGL,		
DTP systems	ASCII		
Alias Full Color			
Show Case			
Estimating tools	ASCII	<------------	

Notes

1.   AutoCAD, the ubiquitous general-purpose drafting package, complements rather than competes with Sonata. AutoCAD users who install Sonata can continue to develop elements and details on their AutoCAD system which can be passed into Sonata in 2D/3D DXF format for building model assembly, coordination and visualization.

2.   Facet is an extensive range of M & E analysis programs running on PCs, UNIX workstations, including HP-Apollo, IBM RS6000 and Silicon Graphics. A direct, dynamic link between Facet and Sonata has been developed. A formal agreement exists between Alias and Facet for marketing the Sonata - Facet solution to M & E engineers in the UK and through distributor channels worldwide. Facet is available in metric measurements only and certain programs would not be appropriate in some countries - please refer to Alias - Facet Marketing Guide to be issued August '92.

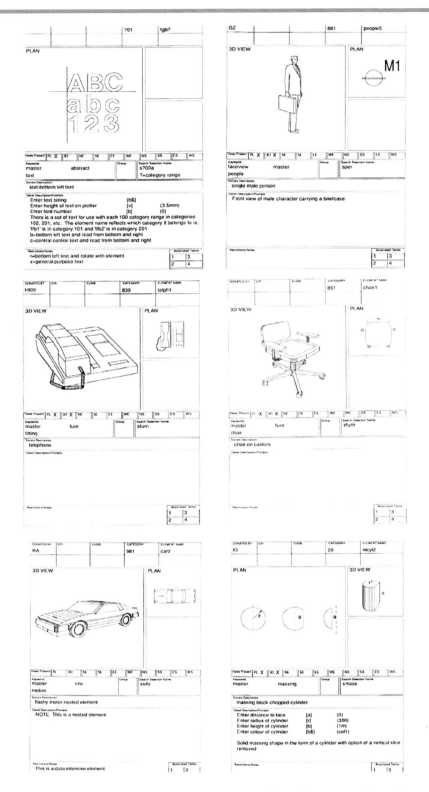

Figure 252. *Sample components SONATA Master Library © SONATA Software Systems Inc*

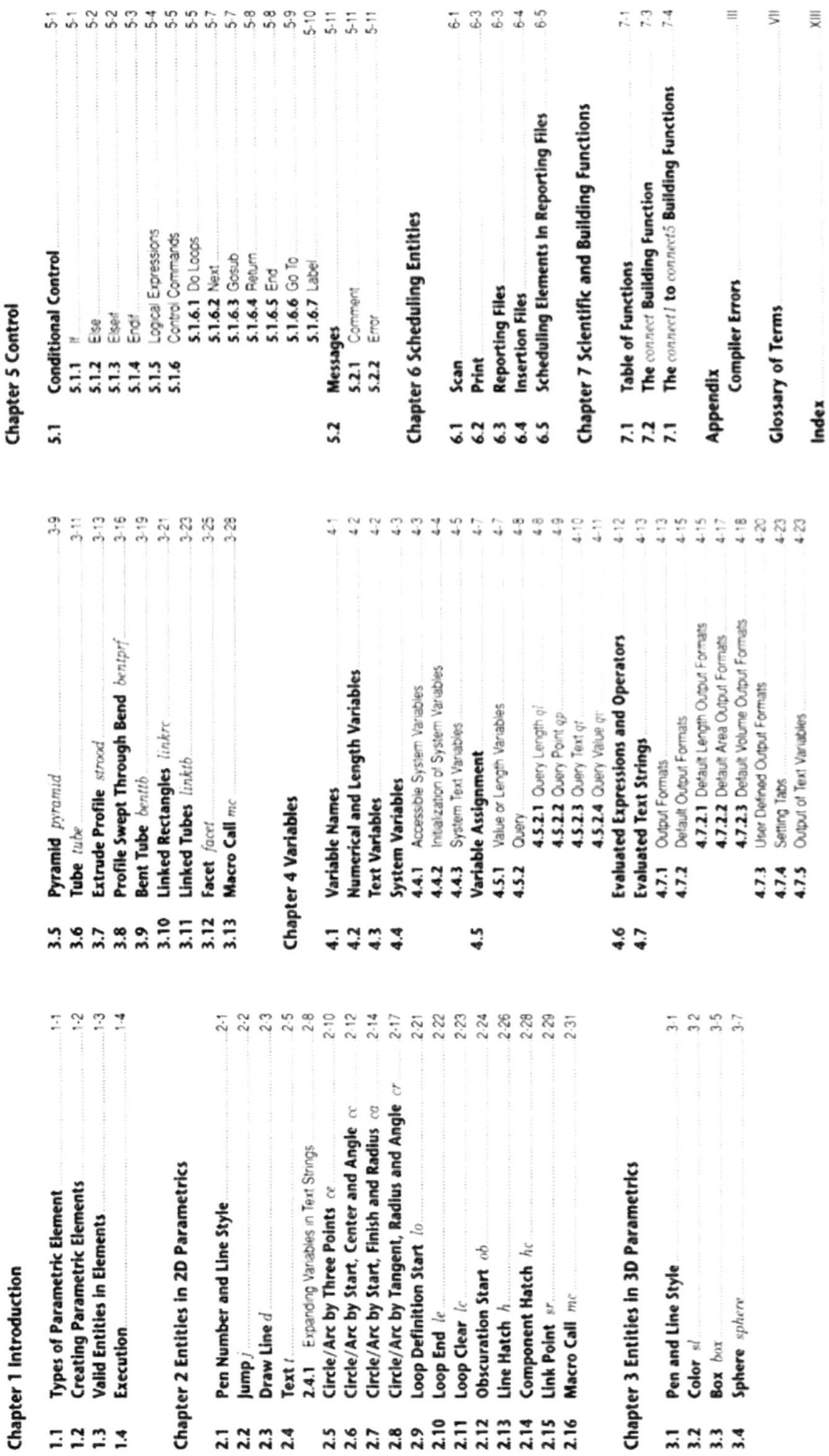

Chapter 1 Introduction

Chapter 2 Entities in 2D Parametrics

Chapter 3 Entities in 3D Parametrics

Chapter 4 Variables

Chapter 5 Control

Chapter 6 Scheduling Entities

Chapter 7 Scientific and Building Functions

Appendix

Figure 253. *Table of Contents of SONATA Parametric Manual © SONATA Software Systems Inc*

Sonata Mechanical and Electrical Library

❑ **Scale Dependency**

All elements have the ability to vary the level of detail shown at different scales. Which means that you don't need to duplicate drawings for different scales.

❑ **Quality Assurance**

Elements have the ability to give visual and audible warnings if they are used outside preset rules. This helps to reduce "bad practice" on your working drawings.

❑ **Clash Detection**

All elements may be used to find clashes between services or services and structure.

❑ **Link Points**

Many elements have link points that enable the exchange of parameters between them.

In designing ductwork for instance a section of duct will automatically size itself to the section or fitting it is being connected to.

❑ **Hidden Lines**

Drawings of complex services have hidden lines automatically removed or are shown dotted where services cross. A major productivity benefit when changes are made.

The library is currently only available with metric sizes.

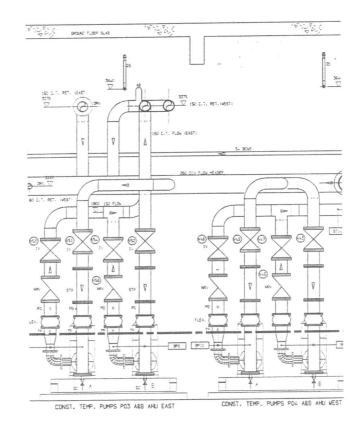

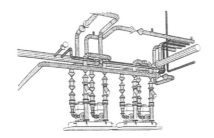

Alias

Alias Sonata Limited
Prince Edward Street
Berkhamsted
Hertfordshire HP4 3AY

Telephone (0442) 865481
Facsimile (0442) 872670

Figure 254. *Description of SONATA Mechanical and Electrical Library © SONATA Software Systems*

Sonata Mechanical and Electrical Library

A comprehensive Sonata element library containing many of the most commonly used building services elements including: ductwork, pipework, electrical wiring and fittings, and drainage.

The Sonata mechanical and electrical services library will prove essential to engineers and technicians using Sonata for both small and large scale building services projects.

Features

❏ Ductwork

 Includes rectangular, circular, flat oval ducts and associated fittings with bends, branches, tees etc.

❏ Pipework

 Includes steel, galvanized steel and copper pipework with associated fittings.

❏ Electrical

 Cable trays and trunking including fittings, luminaries, switching power and distribution elements.

❏ Drainage

 Includes horizontal and vertical falls from 50mm to 300mm with branches, bends, tees, etc.

❏ Schedules

 All elements can be scheduled to give accurate material take-off's from any area of the building. Schedules can also be passed to other systems for such applications as estimating or buying.

❏ Flexibility

 Any element can be modified if necessary to meet your current working practices. New elements can be added at any time to keep your library completely up-to-date with new working standards.

❏ Views

 All mechanical elements include 3D views and, where necessary, schematic views in section, plan, elevation and perspective. This gives you the ability to automatically cut "working" drawings at any point in the building.

Alias
Sonata
the new movement in
Computer Aided Design

Sonata Desktop
Sonata Designer
Sonata Concept
Sonata Concept+
Sonata Concept++
Sonata Fly-through
Sonata teamNet
Sonata Scheduler
Sonata Parametric Element Editor
Sonata Digitizer
Sonata Mechanical and Electrical Library

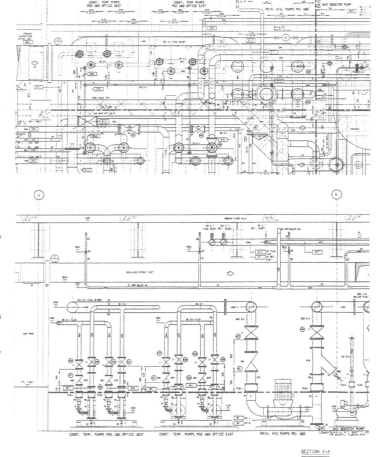

Figure 255. *Description of SONATA Mechanical and Electrical Library © SONATA Software Systems*

SONATA
SEES THINGS
THE WAY YOU DO.

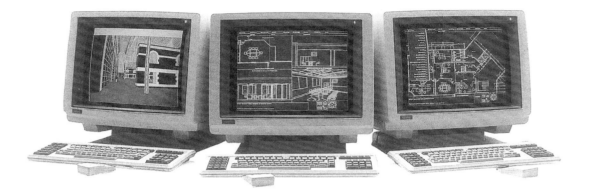

Sonata is today's most advanced CAD system... a generation ahead of the rest.

When we launched Sonata you may remember our advertisement "It makes the other CADS look like bounders".

But a recent purchaser put it even better. "Sonata sees buildings the way we do... sometimes in 2D and sometimes in full 3D but always totally coordinated one with the other."

While another chose Sonata because: "The colour images make our designs come alive in front of our clients".

A financial director who has seen Sonata in action says, "The drafting speed and design coordination make an unarguable financial case for Sonata."

And as Building Design Magazine commented on Sonata's recent showing at the CIC Exhibition at the Barbican: "The biggest splash at the exhibition was the arrival of Sonata".

High praise. But deserved.

Sonata's extraordinary ease of use, low price, first class support and training services could only come from t² Solutions who have enjoyed 10 years of CAD leadership with their famous Rucaps systems.

Sonata looks set to strengthen that leadership for the next 10 years.

See a live demonstration of Sonata at one of t²'s Investment Workshops or ask for a copy of the Sonata demonstration videotape by telephoning Phillipa Jowett on 04427 5481.

You'll soon see why Sonata sees things the way you do.

the new movement in
Computer Aided Design

t² Solutions Ltd, The Teamwork Centre, Prince Edward Street, Berkhamsted, Herts. HP4 3AY.
Tel: (04427) 5481 Fax: (04427) 72670.

Figure 256. *SONATA Advertisement 1988*

Appendix 3

Videos (get the UnderstandingBIM app in your app store)

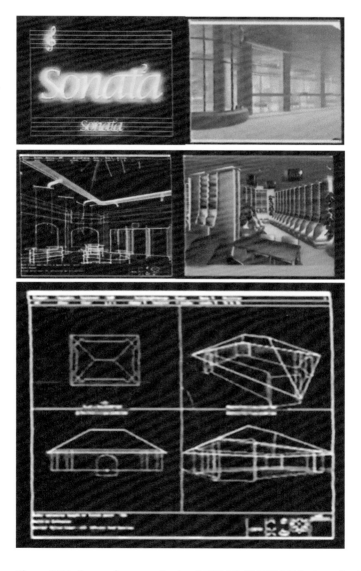

Figure 257. *Scenes from movie about SONATA © SONATA Systems Inc 1988*

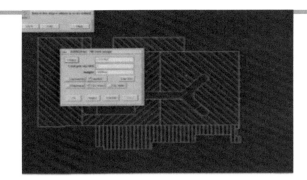

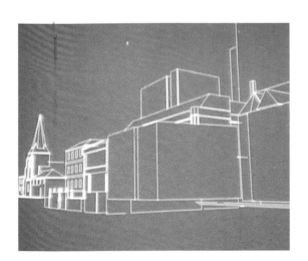

Figure 258. *Project Modelling Movie © Taylor Woodrow 1998*

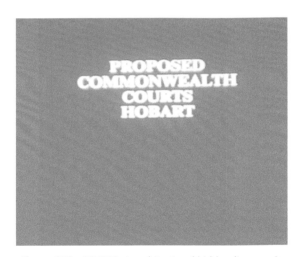

Figure 259. *1977 First architectural hidden-line movie*

Figure 260. REfLEX Movie Melbourne of the City Centre Soft Shadows © e-Arch 1998

Figure 261. SteelEel renewable power from the waves

Appendix 4
SONATA "Plan" 1984

This document, written by the Author in 1984, describes the functionality of the new system. This, and the document in the following chapter, describe the functionality implemented by the Author in SONATA They describe a full BIM system. These

2.   INTRODUCTION

Design Technologies is developing and will market a turnkey computer aided design system for architects, engineers and other professionals involved in planning, designing and constructing buildings. The chosen focus is the Building Industry segment of the CAD market.

This system is unique in the computer aided design field in a number of ways.

1.   It uses a single model to fully co-ordinate graphical and non-graphical information of all objects within the building.

2.   Sophisticated programs allow extremely large models to be built, manipulated and analysed.

3.   Low price entry based on the most powerful micro-computers available today, easily networked within the office.

4.   "Intelligent" components, able to design and modify themselves depending upon the surrounding conditions in the model.

5.   Easy implementation of a small silent desktop unit without special environment requirements.

6.   Simple to use systems which can be used by "non-computer" personnel.

The standard workstation allows input, modification and plotting of construction drawings and perspective views of building models.

Options include:-

(a)   A system to allow viewing of colour-perspective images with shadows, transparent objects, textures in various lighting conditions both internal and external.

(b)   Word processing and payroll packages integrated.

(c)   Integrated relational database access to allow extraction of schedules, costing, invoices etc.

Other possibilities which may be implemented include network design facilities, structural steel design, thermal analysis and so on. (Note that (b) (c) and the other possibilities would all be third party software.)

- 4 -

3.   MARKET POTENTIAL

U.K. Market

There are about 1500* CAD installations currently in the U.K. some 300 of these being in the Building/Construction Industry. About one half of these lie within the vertical segment of architecture. The investment of the building industry in CAD is around £35* million (*Draughting and Design, 21st November 1984).

The main contenders for this market are GMWC and ARC, GMWC having about 60% of the market. These two forces will become less significant during the next few years for the following reasons:

*   RUCAPS, the product of GMWC is some 10 years old on the downward part of its product cycle and extremely difficult to enhance. GMWC have not yet begun to replace this product and have lost key personnel concerned with their ability to produce a new product.

*   ARC has recently been taken over by MacDonnell Douglas, the U.S. aircraft manufacturer. As a result, the emphasis of ARC is believed to be moving away from the architectural segment to mechanical systems.

Within the vertical market of architecture, the majority of systems have been, in general, in the top 10% of the market. This is because of the pricing structure of current systems. For example, GMWCs system price starts around £60,000 and the average installation price is over £100,000.

The market segment being attacked by Design Technologies is the mid-range of the architectural market away from the larger competitors and yet having a greater perceived quality and functionality. This mid-range market is virtually untapped and with some 8000* practices in this country alone, most of which could afford a CAD system, the potential market is large. Penetration of just a few percent a year could be extremely profitable.

- 5 -

The Model

Buildings are designed as number of layers, each layer being assigned a name and description and a datum height. The layer might, for instance, consist of all electrical services on the 33rd floor of a building. The model consists of the positional information of each object, this being entered and altered easily through the graphics interface.

Object Definitions

Each object in the building can be made up of a number of views, usually a plan, several elevations and a three dimensional model. The three dimensional model can be used to create the plan and elevations or vice-versa.

The simplest objects can be input simply by tracing out shapes. From these simple objects through to the intelligent objects is a continuous spectrum. The most complex objects might have many different rules and much knowledge built into them allowing the user great power in their use within the building. For instance, a staircase object might have structural design formulae, rules concerned with the flow of people according to the relevant British Standard and some knowledge of materials. It could interrogate the building itself about the inter-floor height etc.

Objects can be "sketched" in initially and later in the design be replaced by more complete definitions.

Building Construction

Objects can be placed anywhere in the building using very simple graphical means. The grids reflecting the layout of the building can be defined by the user to assist with positioning objects both horizontally and vertically. The current state of the building and the required layers can be viewed on the screen and providing the user has access rights to those layers, the model can be changed at will.

- 10 -

Drawing definitions can be defined graphically and stored for repeated use. These are also accessed by name. Both the construction and perspective drawings references can be generated by this simple mechanism.

Colour Option

A full colour option is available which allows the user to generate colour perspective images of the model. By specifying the latitude and longitude of the site and the time and date, the sun position is calculated automatically allowing accurate lighting and shadow effects to be observed. Internal lighting will be positioned graphically (as objects) with the lighting intensity, colour direction being constructed as part of the object itself. Colouring objects can be done interactively by allowing the architect/user to move through the continuous colour range of 16 million possible colours.

The ability to design in colour with the different lighting effects allows an effective presentation and design tool, enabling the client to experience the building before construction.

Management, Schedules, Financial and Inventory Control

Much non-graphical information is associated with the design and construction of a building. A number of third party software products are included as options with the D.T. system.

* Word processing

    To allow the office to make fullest use of the machine, to allow production of specifications, to allow lengthy annotation of drawings and permit maintenance of the schedules.

* Financial packages

    To assist in pay, forecasting, cash flow, profit and loss etc.

- 11 -

267

*   Scheduling packages

    To allow production of schedules of items, costs and quantities within the building.

*   Management

    Query and report handling for management applications. These could include project progress to date, labour hours, drawing set definitions, project-machine personnel usage, production schedules, accounts payable and receivable etc.

*   Inventory Control

    A complete inventory of furnishings, fixtures and facilities accumulated during each phase can be passed to the facilities management organisation upon building occupancy. Purchase orders can be produced automatically for interior designers and facilities planners.

The product consists of the computer and accompanying software, the software representing the majority of the value added.

Hardware

*   Inexpensive micro-based computer running under the defacto standard UNIX.

*   Small, silent, desktop operation, no special environment required.

*   Powerful workstation for each desk. Power equivalent to £30-£70 thousand minicomputer (VAX 11/750 or Prime 2550).

*   Standard high speed network interface available for linking machines and other devices.

*   Integrated high resolution graphics screen and mouse for high speed interaction.

- 12 -

* Standard IBM interface available.

* Integral disk drive up to 48 Megabytes.

* Virtual memory (no restriction on program size).

* Optional colour screen via high speed interface.

Software

* Written in standard FORTRAN 77 with UNIX operating system and GKS graphics interface allowing easy porting between machines.

* Carefully designed and structured to modern standards to ensure minimum maintenance and easy up gradability.

* Fully documented for both the user and the software analyst.

* Ergonomically designed user interface to allow high speed interaction, simplicity of use and a self teach mode. Designed in close conjunction with architects and engineers.

* Innovative in many aspects, in particular:

    high speed database design
    "intelligent" rule based components
    sophisticated 3-dimensional visualisation
    integrated 3rd party software
    full 2-D functionality

- 13 -

Appendix 5
SONATA Implementation 1985

Written some two years later in December 1986, this document describes the actual implementation achieved by the Author. This system meets todays definition of BIM as defined by the BIM Handbook.[1] These documents were used in fundraising at the time. (Note: The image on the second page has been lost.) The alterations to the document were made at the time reflecting ongoing development.

This document is now part of the RIBA Archive at the Victoria and Albert Museum in London and, as such, is © RIBA Library Drawings & Archive Collection, Victoria & Albert Museum.

---

1   BIM handbook Eastman et al Wiley 2011

THE DESIGN TECHNOLOGIES SOLUTION

DECEMBER 1985

INTRODUCTION

The Design Technologies Solution to the CAD problem represents the next generation of CAD systems for the Construction Industry. It embodies the most effective concepts of existing CAD systems together with a number of important innovations, allowing a "cradle to grave" philosophy in the design process.

Some of the features include:

* Simplicity of use achieved through the popular icons, windows and pull down menus in conjunction with a clear user conceptual model.

* Object based forcing coordination between various drawing types and non graphics data as well as facilitating fast construction and revision of building models.

* "Intelligent" objects capable of self-design and self-detailing and possibly the implementation of some Regulatory Codes. Various network and connectivity calculations can be solved by utilizing connectivity capabilities between objects.

* Integrated non-graphics facilities which can generate reports, schedules, orders and other non-graphics information.

* Comprehensive display functions including hidden-line and colour perspective generation.

The new system is the result of careful analysis over a number of years of existing CAD systems by experienced CAD users and designers.

In its simplest form the new system is straightforward to use, leading the user through a continuum of experience to produce complex models with comprehensive self design capabilities. Broad conceptual buildings can be generated quickly being refined and changed as the design process progresses.

REVIEW OF CURRENT SYSTEMS

CAD systems come in many shapes and forms; some try to be all things to all people, others provide such minimal capabilities as to be unusable. Experts alone can use the more complex systems whereas others are so expensive that there are never enough work stations for any particular design office. Some fill a very small part of the design process so as to be not available, others are clumsy resulting in user frustration, and still others have many separate processes with separate data entry for each. All of these systems are capable of producing lines on paper. The concepts for achieving the total design solution are those in question.

General purpose CAD systems suffer from a number of disabilities when used in the Construction Industry.

* They are usually designed to model machine parts and complex surfaces. Constructing a building with such a system can be difficult.

* Performance degrades rapidly when thousands of instances of objects occur in a single building.

* Terminology is not familiar to the Construction designer.

* Many features, which are not useful, clutter the already crowded menus.

The new system is aimed at the design of large buildings and factories. Features and terminology are tailored for this group and the database is specifically designed to deal with the largest models.

CAD systems used in the Construction Industry can be classified as two dimensional, three dimensional and a combination of these.

2D systems represent building projects as a series of 2D layers which combine together to form drawing sheets, lines, arcs, text, etc, which are stored in these layers and can be combined to form cells or groups. This type of system allows the drawing board to be replaced in the current design process. This means that the problem of reconciling the drawing with the objects represented is always left to the draftsperson. No coordination is maintained between the various views of a building allowing fundamental differences to occur between, say, plan and elevation views. In those 2D systems where non-graphics capabilities such as scheduling are available, the results can differ depending upon which drawing is analysed.

3D systems model buildings as 3D geometric objects. They can generate perspective and sectional views of a model, but fail where different views require symbolic representations. For instance, for a window the symbolic elevation and plan views and perspective view are quite different and cannot be reconciled geometrically. Another problem is that 3D objects are more difficult to generate than the 2D symbolic equivalent. Given that many projects are not viewed in perspective, the designer is forced to do extra work.

A number of systems combine 2D and 3D ideas. The 2D world is usually separated from the 3D world. When either the 2D or 3D world is changed (depending upon the system) the other must be generated. Symbolic information, notes and dimensions must be stripped out or reinserted depending upon the system.

The new system permits a specific building object to be represented both as a 3D geometric view and a group of 2D symbolic views. Building a single model with such representations allows full coordination to be achieved between the various views and allows symbolic and perspective views to be generated. If an object representation or position within the model is changed then all subsequent drawings reflect this change.

Many existing CAD systems suffer from inconsistent user conceptual models, numerous obscure programs which must be run to perform individual processes, poor user interfaces, inadequate prompting and over ergineering. The amount of training for such systems is considerable, resulting in long lead times before the system is productive. Similarly, firms are at the mercy of the few designers sufficiently skilled to use such systems.

By using the highly successful and easily understood icon/window approach developed at Xerox, and with careful prompting new users can produce drawings within hours. These simple graphics lead naturally into the more sophisticated commands. Having multiple desktop machines means that users can learn at their own pace without a queue of people waiting. Non-graphics data extraction is no longer a separate set of commands and language but simply a variant of the normal objects he places within the building.

The traditional focus of giving the designer a tool to work faster does not go far enough. By creating the capacity to offer new services as part of the total design solution, the firm can deliver not only new applications but also a better product through an improved design process.

* Better design solutions based on the ability to review more alternatives visually, environmentally and structurally.

* Acceleration of delivery of the project through improved project control, reduced on-site conflicts, automatic compliance with Regulatory Codes.

* More accurate projection of costs through bill of materials estimating and automatic design and detailing.

The designer using the new system could also claim benefits as he is able to reach new groups of clients, thus diversifying into the markets. Examples could include: facilities management services, urban planning through visualization, space planning, comprehensive services based on structural and environmental analysis.

Further benefits could be achieved by expanding the firm's capability without adding staff and larger projects could be taken on.

THE DESIGN TECHNOLOGIES SOLUTION

The more important features of the new system are considered. These are listed in point form for brevity.

* Speed. Access to any part of the building, is fast even for the largest models using a specially developed hierarchical database.

* Data Management. Methods are provided for selecting groups of objects based on keyword, categories, dates and wild card letters in the name.

* Model accuracy. 32 bit accuracy is maintained throughout the model. Care is taken in performing floating point calculations to ensure accuracy.

* User defined grids. Users can construct and lock onto the most complex and hence building shapes.

* Named layers at various heights within buildings. Various layers of each floor level based on functionality. Access to change such layers is based on optional passwords.

* Multiple degrees of freedom for components in layers.

* No restriction on type or amount of information associated with components placed in layers.

* User defined grids enabling odd shaped buildings to be assembled.

* High 2D functionality including splines, user defined line styles, hatching and $2\frac{1}{2}$ D obscuration. Multiple symbolic and 3D views can be associated with each component.

* Input and output dialogues when components placed, building allows all components to request information and output intermediate results. Defaults and an "unset" mechanism means not all values need to be set when placing a component.

Three Dimensional Capabilities

A number of powerful 3D capabilities have been included.

* 3D component generation programs using standard shapes and surfaces of revolution and extrusion are combined using boolean operators.

* Mixing of 2D and 3D data and plan/elevation stage and in perspective views.

* Computational facilities as in 2D components.

* Fast hidden surface output and a slower more realistic output method giving shadows, multiple light sources and spot lights.

Conclusion

Designers new to the system can produce drawings quickly and easily. As skills increase and requirements change powerful facilities for manipulation and control of the database become available. Components can be created to perform many different design and detailing calculations. Non-graphics information can be extracted and formatted through special components.

It is intended that this system be used as a single workstation in the smallest design office and yet also be found in large firms, networked sharing a single database.

SYSTEM DESIGN AND REQUIREMENTS

The complete system is written in structured Fortran 77. This was done to ensure portability to enable the target machine to be selected as late as possible. As part of the portability push as much code as possible has been written in 32 bit integer. The places where floating point calculations are used global tolerancing has been used to ensure relatively painless porting. Although implemented in a Unix environment very few routines are operating-system dependent and can be found in a single library (dates, time, shift functions, accounting and some file handling). The graphics requirements are basic in that text is handled internally and the segment handler and multiple window manager are written in Fortran. Several modules from these libraries could be written in machine code to greatly enhance performance. Handles to pull down menus, window panel handling, icon handling and cursor control are required.

The code consists of some 50,000 lines of code (60% comments) in about 700 subroutines. Memory requirements are depending upon size of models to be viewed (with hidden lines/surfaces removed) in 3D.

The principal computational sectors are divided into separate stand alone libraries for easy reuse and maintenance. All subroutines are "self documentary" to the point that all have keywords in their headers to allow automatic extraction of descriptions, argument list and description, and modification dates.

Appendix 6
Letter to the Author

This is the text of a letter sent in 2016 to the Author from Gábor Bojár, the founder of Archicad. Gábor has kindly agreed to allow the publication of this text. Further relevant correspondence might be found in the Understanding BIM app.,

"We always highly respected and regarded Sonata as one of the best – if not the best – competitor of ours in the early days of our market entry in the UK. We particularly admired that this excellent piece of software was written by a single person, when all other competing solutions (including our one) has been built by large development teams. In addition, we agree that from quite a few technical point of view Sonata was more advanced in 1986 than ArchiCAD at that time.

We still believe, however, that the first release of ArchiCAD in 1984 (called RadarCH on Apple Lisa) with its 3D building modelling capabilities, parametric and intelligent 3D objects, can be regarded as a pioneering predecessor of BIM. Having said that we agree that in 1986 Sonata surpassed already the matured definition of "BIM", specified only about one and a half decades later.

Regarding the influence of Sonata on the ArchiCAD development, it naturally influenced us, without infringing any IP of Sonata or T2. Competitors always do study and influence each other, this is a standard and fair business practice, as long as competitors refrain from infringing each others."

Index

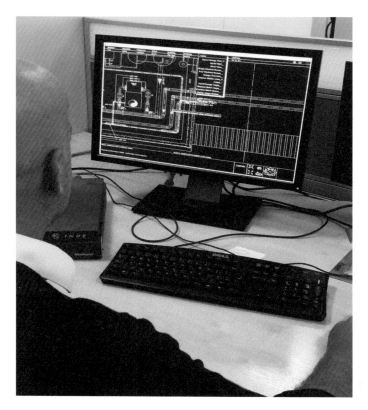

Figure 262. *Father and Son @ BAM Construct UK 2015*

Figure 263. *The Prince Phillip Gold Medal awarded by the Royal Academy of Engineers to the Author in 2016*

T - #0640 - 071024 - C314 - 276/219/18 - PB - 9780367244187 - Gloss Lamination